BACKYARD FARMING

➤ *Make your home a homestead* ◄

☞ GROWING VEGETABLES & HERBS

"EXPERT ADVICE MADE EASY"

Kim Pezza

🌊 **hatherleigh**

🔊 hatherleigh

Hatherleigh Press is committed to preserving and protecting the natural resources of the earth. Environmentally responsible and sustainable practices are embraced within the company's mission statement.

Visit us at www.hatherleighpress.com and register online for free offers, discounts, special events, and more.

Backyard Farming: Growing Vegetables and Herbs
Text copyright © 2013 Hatherleigh Press

Library of Congress Cataloging-in-Publication Data is available upon request.
ISBN: 978-1-57826-459-9

Cover Design by DcDesign
Interior Design by Nick Macagnone

Printed in the United States

10 9 8 7 6 5 4 3 2 1

www.hatherleighpress.com

TABLE OF CONTENTS

INTRODUCTION

Gardening has been an important part of life throughout the centuries, not only in the United States but across the world. However, in recent history, vegetable gardening has disappeared from many home landscapes as cities, suburbs, and even rural areas have turned to the local grocer for their vegetables. The reasons for this shift vary, but it is often a result of busier lifestyles and the need for easier access to fresh produce.

But in the last few years, a resurgence of local farming and a renewed consumer interest in organic, homegrown produce has revitalized home gardening as a pastime and as a food source. Consumers have become more interested in where their food came from, passing over genetically modified foods and mass-produced produce in favor of organic, locally grown fruits and vegetables.

Individuals, families, and even restaurants are beginning to find their way back to the land. As a result, through local farmers, farmer's markets, and backyard gardens, America is experiencing a food revolution that shows no sign of letting up. These are the foods that our parents, grandparents, and even ourselves as children grew up eating and thriving on. Consumers today are looking to gain back control of their food by returning to farm-fresh produce. Many are also becoming interested in growing some of their own foods as well. As demand continues to grow, more and more individuals are looking to take up the spade and trowel and try their hand at providing their own farm-fresh produce.

Growing Vegetables and Herbs is ideal for those of us who want to join the move back to basics, especially those who have never planted a seed, picked a tomato off the vine, or fought with a weed…and had it win! Covering methods that can fit many situations, including limited gardening space, *Growing Vegetables and Herbs* will help you create your own gardens filled with vegetables and herbs, as well as some fruits and edible flowers.

Everything you need to begin your own gardening experience is contained in these pages. Read it from cover to cover or read it section by section as new questions and challenges arise. Either way, this book was written with the first-time gardener in mind, and is specifically designed to equip you with everything you need to make your first attempt at gardening fun and simple.

You will find nothing more rewarding than tasting foods you grew yourself. Imagine sauce from your own tomatoes, fries from your own potatoes, or pickles from your own cucumbers. Not to mention the dishes you can create with the freshest ingredients possible, direct from your own backyard! Once you taste freshly made foods from your own kitchen using produce from your own garden, prepackaged or grocery-store food will never measure up again.

So sit back, kick up your feet, have your favorite beverage at your fingertips, and begin your gardening journey. Whether you have one hundred acres, live in the suburbs, or are a full-time apartment dweller, you're on your way to creating your perfect garden.

MEET THE EXPERT

Kim Pezza grew up among orchards and dairy and beef farms having lived most of her life in the Finger Lakes region of New York State. She has raised pigs, poultry and game birds, rabbits and goats, and is experienced in growing herbs and vegetables. In her spare time, Kim also teaches workshops in a variety of areas, from art and simple computers for seniors, to making herb butter, oils, and vinegars. She continues to learn new techniques and skills and is currently looking to turn her grandparents' 1800s farm into a small, working homestead.

CHAPTER 1

THE BASICS

I n most cases, the new homesteader, backyard or urban farmer will begin their journey with a garden. It may be very large or very small, raised beds or containers on a balcony, in a backyard in the country or on a rooftop in the city. No matter what type of garden it is, for the first-time gardener, it can be one of the most gratifying projects someone can start.

In all cases, no matter the size or location of your garden, it is a food source, not just landscape aesthetics. This isn't to say that a vegetable garden cannot also be part of your landscape or even the majority of your landscape. Even a flower garden can have a purpose other than just being decorative. The flowers may be edibles or fruits and vegetables may be intertwined within the decorative flower garden (of course, they may intertwine with edible flowers as well).

Depending on your family's needs, wants, and available space, the garden may be a supplemental source of food. However, in some cases the garden may also be part of a total self-sufficiency plan. Some gardens may also be grown as market gardens, meaning gardens grown expressly for selling the harvest, while others may be grown mainly for personal consumption, but any excess will be sold (after you have eaten, canned, frozen, or dried all you want for the family), usually selling from a little stand in the front yard.

Whatever your garden is being grown for, you will need to think about its development. This does not necessarily mean that you need to take the time to create a map or a scale drawing, but site, size, and what the plot will be able to hold does need to be thought out, as well as what type of garden will be used. In Chapter 2, you will find more information on how to plan your very own vegetable garden to suit your needs.

Today, food gardens can be found in a variety of locations besides the farm. Backyards, empty city lots, schools, rooftops, businesses, apartment buildings, and even balconies can be found making a home for some type of garden. Basically, whether there are multiple acres or a couple of pots on the porch, food may be grown. And although there has been much ado about a vegetable garden in the White House, this is actually nothing new.

History of Vegetable Gardens

Starting in 1800, John and Abigail Adams had a family garden during their time at the White House. Then, in 1801, Thomas Jefferson, a noted farmer in his private life, added fruit trees to the White House gardens. Upon his arrival, Andrew Jackson included his touch with the addition of a small greenhouse, which was later replaced by a full-size version (unfortunately, in 1902, it was torn down and replaced by the West Wing).

In addition to food gardens, some of the First Families throughout history also kept livestock on the White House grounds, making this most famous house no stranger to the backyard farm. While today, the food garden seems to be a trend going back to early American life, it has been found that people have most likely been gardening for at least 8,000 years, beginning in an area of the world called the Fertile Crescent.

The Fertile Crescent is a very ancient area of rivers that stretch from the Nile to the Tigris and Euphrates, which features fertile soil that also provided pasture land and a wonderful growing area. Part of the Biblical lands area, it was the home of mainly nomads, who needed the good pasture lands for their livestock. However, the land was such a perfect agricultural area, due to the natural fertility and the presence of the rivers that allowed for irrigation, the nomads settled down in the area. It is also believed that the Fertile Crescent is the location of the earliest known culture.

The earliest known vegetable gardening book is actually still available in the marketplace. Written in 1599 by Richard Gardiner in Shrewsbury, England, the book focuses on such vegetables as leeks, turnips, lettuce, squash, and beans. The title of this publication is *Profitable Instructions for the Manuring, Sowing, and Planting of Kitchen Gardens Very Profitable for the Commonwealth and Greatly the Helpe and Comfort of Poore People.* Quite a title, even for the sixteenth and turn of the seventeenth centuries.

In the early United States, it was the kitchen garden that was the household staple, with the specific purpose of feeding the family, sometimes being the only source of food during lean hunting times. Getting its name due to its proximity to the kitchen, the kitchen garden held the seeds that the pioneers or immigrants brought with them from the homes they had left behind. And although it is not quite the necessity that it was in early America, the kitchen garden has been making a comeback, supplying vegetables and herbs to the gardeners of today.

Benefits of Growing Your Own Food

Although many still think that a big garden area is necessary to grow vegetables or fruit, this cannot be further from the truth. You don't need acres of land to start a vegetable garden. Even with a few raised beds or a terrace garden, you and your family can enjoy the many benefits of growing your own food.

When you grow your own food, you have total control over what goes into your cupboards and what you eat. You also have total say as to how the food is grown, how it is harvested, and how it is preserved or prepared. You can even have a good stock of homegrown foods to keep you through the winter, should you decide to can or freeze the excess. You'll find that by February, you will appreciate those canned tomatoes or the soups that you made last fall.

Another benefit of growing your own food is that it teaches your children about where their food comes from. Learning about and even helping with the growing of their own food helps children to develop good working habits, as well as an understanding of responsibility. It has even been shown that community gardens can actually spur a total neighborhood cleanup, due to the unexpected pride gardeners find they have for their hard (but gratifying) work.

For those who have allergies, growing your own food will ensure that the food you are eating has not been grown or processed with something that could set off a reaction.

And finally, you can grow exactly what your family likes or loves. And if there's something you like, but can't find in the grocery store, there is a good chance that you can grow it yourself! (But remember, your climate may have a say in what you can put in your garden.)

While there are many reasons to grow your own food, the best reason is the satisfaction you will take from growing foods for your own family. Whatever the reason, whatever the need, you will find a garden to be nothing but a wonderful experience, and one that you will look forward to year after year.

CHAPTER 2

PLANNING YOUR GARDEN

. . . .

As a new gardener, a fun activity can be to put together a wish list of your goals and expectations for your perfect garden. As you do this, it is important to also be realistic about what will be feasible for you. It can take some thought, planning, and sometimes a little bit of compromise to build that idea into the garden of your dreams, but the planning process can be a wonderful learning experience.

Here is an example of some things to consider when putting together your wish list:

- How large a garden do you want? Remember: if you are limited in space, your options may be limited in this aspect.
- What do you want to grow, and how much? (This can always be adjusted year to year.)
- Are you just growing for yourself, or do you hope to sell your produce as well? (Again, this may depend on space available.)
- Do you want to grow organic? If so, do you have all the resources you will need (organic seeds, fertilizers, etc.)?
- If you have the space, will you have one large garden or a number of smaller ones? Sometimes making a few smaller gardens instead of one big one can have a positive impact on the look of the landscape, without affecting what you can grow.

- Do you want a simple garden or a more elaborate one? Just remember: the more elaborate, the more time (and money) it may take. A simple garden can produce just as well (and as much) as an elaborate, highly decorative one, and may even be a bit easier on the wallet and schedule. However, should you decide later on that you do want something a bit fancier, you can still do so and make the necessary changes at any time.

- Will you be doing **companion planting**? If so, then you will need to keep in mind which plants will do better together and which plants will need to stay clear of each other. You will need to take this information into consideration when planning your garden, even if using the container method.

- What shape do you want your garden to be? If you are building a traditional or raised bed garden, you can make it in almost any shape you want. The usual shapes are square and rectangular, but some are round or oval, while some raised beds may even be tiered (if you have the time to spend).

- Will you have small children working in the garden? If so, you may want to put in a small section of the garden that would be easier for them to work in, using plants like lettuce and cherry tomatoes, which will be easy for them to tend and harvest. If you give them their own small section or even their own small garden to tend to, then the adults and older kids can work their gardens without the little ones wanting to "help," which can sometimes create problems. If you are growing for market, raising highly unusual foods, or any other situation in which you cannot afford an accident by little hands, it's best to keep them out from underfoot.

These are only a few things to keep in mind when creating your own wish list. Your list will of course reflect your own wants and needs.

Determining Type, Space, and Size of Your Garden

Container gardening, vertical gardening, and even small raised beds can grow enough vegetables to supplement a family's needs, with container gardening being especially useful for the gardening apartment dweller who would have little or no yard space for a traditional or raised bed garden. It should be noted, however, that container gardening may not produce quite as much as a small traditional, vertical, or raised bed.

So, when creating a vegetable garden today, how do you know what size and type of garden to create? This is easy enough to answer. Since the size of your garden will be limited to your available space, the first thing to do is look at what is available as far as this space. Then you need to look at how much time you have to commit to the garden, as well as what is to be grown.

If you don't have a lot of time to devote to a garden, a container garden may work best, no matter how much space you have available. If you live in an area where drainage or soil is not good, then raised beds may be the answer (see page 19 for more information on these topics). Whatever you decide, should you want to change course the following year, you can (provided you have the space). Don't assume that the garden you choose this year will be what you're stuck with next.

Another thing to consider is what you want to grow. For example, if you want to sell lots of pumpkins for Halloween, then a little two-by-three-foot garden of pumpkin plantings won't cut it. But if you need just a few pumpkins for a jack-o'-lantern and some holiday pies, the two-by-three-foot garden would be perfect.

Don't forget to consider your own physical abilities. If you have a difficult time bending, then a vertical garden could be for you. In a wheelchair? Then consider raised beds with wide, smooth walkways to allow for your chair to operate. You can also adjust the height of your raised bed to whatever is comfortable for you.

There are a number of things to look at when deciding on the type and size of garden. But saying this, don't make thinking about the size or type of your garden an entire project in itself. Remember, if you make a mistake in this year's garden, you can correct it next year.

Finding the Right Type of Garden

A garden, be it flowers or food, can be as simple or as complex as you like.

For the purpose of this book, we will use the most basic of styles that were discussed earlier:

- **Traditional:** With the traditional garden, the plants or seeds are planted into flat, tilled (or untilled) ground. This is what most people picture when they think of a garden. These are the most inexpensive types of gardens to put in.
- **Raised bed:** Raised beds are exactly what they say: garden beds raised off the ground by inches or by feet. The garden beds are created in wooden frames and are usually built up at least eleven inches off the ground. Raised beds may be in frames built on the ground or in frames that are raised up on legs.
- **Container:** These are simply gardens in pots or other containers. The containers can be flower pots, wooden boxes, bags, or anything else that a plant can be put in for growing. While some plants may have individual needs once planted in a container, if you can pot a flower, you can certainly pot a food plant.
- **Vertical:** Vertical gardens may be either traditional or raised bed gardens. The difference is that everything grows upward. If the plant does not naturally grow upward, it can be trained

to do so by using supports. Supports will need to be used with the vertical gardens, which will allow the vertical growth of the plants.

Any one of these garden styles may be found in backyards, city lots, rooftops, or on rural farms and homesteads throughout not only the United States, but many other parts of the world as well.

Traditional gardening. Photo by OakleyOriginals under the Creative Commons Attribution License 2.0.

Of course, your selection may automatically be narrowed down by what space is available to you. An apartment dweller, for example, will have nowhere near the same space available for a garden as someone with rural acreage or even an urban backyard. And depending on the type of building, there may not even be a rooftop to use. In some cases, it may come down to a small balcony or some bright windows, in which case a container garden would be the only option. But no matter what the space, each will allow some sort of way to grow vegetables, herbs, fruit, or even some edible flowers.

Raised bed gardening. Photo by Lori L. Stalteri under the Creative Commons Attribution License 2.0.

To determine which type of garden you want to create, first put together a wish list of what you want to grow. For example:

- Do you want early or late harvest foods? (Usually, there will be a mix.)
- Would you prefer heirloom or hybrid plants (or perhaps a combination of both)?

• Will you want to do a late-season planting of cooler weather plants? Some vegetables like lettuce, spinach, and similar plants can have a second planting in late summer for an early to mid-fall harvest. Depending on where you live, you may also need to take zoning laws into consideration. Unfortunately, some areas see vegetable gardens as eyesores, and require that they not

Container gardening.

be in street view. So, if most of your available space is in the front yard, your wish list may need to be adjusted to fit the smaller backyard area, out of street view.

Urban Farming

Usually consisting of gardens and sometimes a mini-orchard, urban farming can be anything from containers on a terrace to raised bed community gardens in empty lots. It is very important to check regulations on what is allowed as urban farming, zoning-wise, can be quite restrictive. Selling produce direct from gardens may not be permitted. Tools necessary for the urban farm are primarily common garden tools (shovels, rakes, wheel barrows, etc.), although community gardens may use small tillers to prepare large plots.

Suburban Farming

Usually traditional or raised bed gardens, suburban gardens are commonly found in private yards or community gardens. Although most suburban farms are in nonagricultural areas, there may still be zoning restrictions. Some developments may not allow food gardens in the front yard (some very strict developments may not even allow them in the back). Selling produce from the yard may or may not be permitted. The same tools used in urban farming would be used for suburban farms.

Rural Farming

Rural farms are in the country, usually in agriculture areas. There are usually fewer restrictions in rural areas, though there may be requirements as to lot size if you plan to include animals. There is usually no problem in selling excess produce from a stand in front of the house or starting a farmer's market. Tools are the same as for urban and suburban farms, but as available space in rural farms ranges from small patches to entire acres, some rural farms will have small tractors for tilling and plowing.

Remember, especially if this is your first garden: if some of what you try this year doesn't work as planned, you can rework and make changes for next year. Most garden plans are not written in stone, so don't be afraid to amend your wish list for the following year.

When considering what type of garden you may want to create, also be sure to consult Chapter 3, which provides a detailed overview of each style.

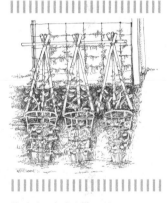

Vertical gardening. Illustration by Ariel Delacroix Dax.

Organic or Not?

Besides space and size, another thing you will need to think about is whether or not the garden will be **organic**. For a garden to be organic, it doesn't matter what type of garden is created. Traditional, raised bed, vertical, and even container gardens may be organic. But it does matter what is added to the soil, which types of seeds are used, and whether you will use pest control. Choosing to go organic means a commitment to keeping your garden chemical free, as well as using only organic seeds and gardening practices.

Although this does not mean fertilizers and pest elimination may not be used, it does mean that whatever is used must be organic approved, especially if your garden is to be certified.

Community Gardens

Community and neighborhood gardens are popping up everywhere, spearheaded by people wanting fresh foods as well as for economic reasons, not to mention many have already had a passion for gardening.

Community gardens are cared for by the people and families of the neighborhoods that the garden is located in, with all who participate sharing the bounty of the harvest. Due to the fact that a community garden would be much bigger than a backyard city garden, and can offer a larger quantity in harvest than a container garden ever could, this garden type could be a preferable alternative for those gardeners in the city who are looking to grow more than what they themselves may have space for at home.

Along with being a wonderful project that brings people in a neighborhood together, community gardens also seem to renew a sense of pride in a neighborhood, spurring other improvements and cleanups in the area as well.

Growing totally organic simply means that no chemicals have been involved in the production of the food. This does not mean that fertilizers and pesticides cannot be used, but they must be a nonchemical and labeled organic or approved for organic use. It may also include the use of companion planting and the use of garden-friendly insects for parasite control.

Although you do not need to purchase organic seeds to grow the plants organically, if you are looking for organic certification, the use of organic seeds would be a requirement, as the foods will not be able to be "certified" unless the food is produced totally organic, which means organic from seed.

Certified Organic

Home gardens may be certified organic, but unless you have a very large garden with a large section of harvest going for market sales, it is not always feasible to become certified due to the high certification fees and record keeping involved. In fact, for most home gardens certification is really an unnecessary step. With that said, if you do intend to eventually expand into a market garden (a garden where harvest is grown for market), because the garden and soil will need to be pesticide free, it would be wise to use caution as to what is used on your garden and possibly begin to follow some of the organic regulations to make the eventual transition easier and faster. For information on organic certification and requirements, check the United States Department of Agriculture Web site at http://www.usda.gov. Most states also have their own organic-certification offices. Local extension offices may also offer information on certification.

Whether to make your garden organic is a matter of personal preference. Many gardeners will take it partway, meaning that they will keep their garden as organic as possible, but may use nonorganic fertilizers and/or nonorganic pest control when necessary. Either method, with proper care, will present a beautiful garden with enough vegetables (and fruit) to make it all worth the effort.

CHAPTER 3

DESIGNING YOUR GARDEN

· · · · · · · · · · · · · · · · · · · ·

Finally, it is time to begin designing your garden. The following sections briefly discuss the four garden types in a bit more detail. Actual planting instructions will follow in Chapter 5.

Traditional Garden

Probably the most recognizable garden type as well as the one that most people picture in their minds when thinking about a garden is the basic **traditional garden** with tilled soil and nice, neat rows of plants. Although usually rectangular in shape, these gardens can actually be any shape or size. Traditional gardens are one of the easiest to install, but they do have their shortcomings.

A traditional backyard garden. Photo by Southern Foodways Alliance under the Creative Commons Attribution License 2.0.

To create a traditional garden, simply till or dig up the ground, remove any rocks and pieces of sod that have been dug up, and plant according to your garden design. Planting can be done in simple rows; however, as you become more experienced, you can create a more designed layout. When planting in a traditional garden, keep companion planting in mind (see page 41 for more details).

With the traditional garden comes the all-important task of keeping up with the weeding. Depending on the size of the garden, this could be a time-consuming task that will need to be performed several times throughout the life of the garden. However, there are a few things that can be done to help keep the weeds down and make the job easier.

After the soil has been tilled and before planting has started, a **garden cloth** may be laid down on the rows to be planted. Garden cloth is simply a very thin material which usually comes in rolls. It is easy to lay and can be a great help with weeding. To plant with a garden cloth, small slits can be cut into the fabric, allowing a small hole to be dug and the plant inserted. Using a garden cloth isn't foolproof, and it usually will not cause total weed annihilation, but the little bit of time it takes to put a garden cloth in place will greatly reduce the amount of time necessary to devote to weeding and can end up being a great help.

For the budget-minded, newspapers can be similarly used as a weed inhibitor. Laid down first as a garden cloth would be, the similarities of installation end here. While the garden cloth lies on top of the garden, the newspaper is laid on the ground, wet thoroughly, and a thin layer of soil is then placed on top of the wet papers. (Although you may want to skip the step of covering the paper, doing so will find newspapers from the garden blowing around the neighborhood during the first good wind. So a little extra time now will save a headache later on.) As with a garden cloth, you would then plant by making holes in the soil and newspaper for each plant.

It should be noted that some gardeners find this method works better than a garden cloth and isn't a budget buster for larger garden plots. However, newspapers may break down after a single season, needing to be laid again the following year, while garden cloth could last two or more seasons. The choice of whether or not to use a ground cover as well as the type to be used will again depend on your time and budget.

Another drawback of the traditional tilled garden, especially if you have limited time to spend with your garden, is the fact that it does not hold moisture as well as the raised bed garden, and may be difficult to use for those with physical limitations, especially those with problems bending or kneeling. However, this is also one of the least expensive gardens to install (unless there is bad soil and heavy improvement is necessary). The traditional tilled garden is great for beginners or anyone who is unsure whether gardening is for them and just wants to try it out first without having to put out a huge investment first.

Raised Bed

In its basic sense, a **raised bed** garden is a traditional garden that has been built within a frame that has been filled with soil rather than being flush to the ground.

The raised bed is a bit more expensive and

Raised bed garden. Photo by rpaterso under the Creative Commons Attribution License 2.0.

time consuming in the initial construction in comparison to the traditional bed. However, these gardens will have much less weeding, hold moisture much better, and because it can be built at nearly any height through various ways of constructing the frame, planting, weeding, and harvesting can be much easier for seniors or the disabled. Because new soil or a soil mix will be added to the frame of the raised bed, even locations with poor soil may still be put to use for a productive garden.

The basic process to build a raised bed garden is simple, although it will take some time, a little elbow grease, and a little bit of money. It is a job one person can do, although if the beds are to be large and many, it will be much easier to have at least two people working on it. The job will go a bit faster too.

Like the traditional tilled garden, a raised bed can be any shape or size, although these beds are usually eleven inches or more in height in order to accommodate enough soil for planting and depending on the desired height. However, before anything, the box frame(s) the garden(s) will be housed in must be built.

Materials used to build the box can be any material that would be for the most part food safe. In other words, if landscape timbers are to be used to create the bed box, and if the wood has a preservative on it, make sure that it is a nontoxic type or else the plants and food they produce could be affected. Otherwise, it is preferable to use woods that have not had a preservative applied. (The same goes with paint. If wood is to be painted, make sure it is nontoxic.)

The reason for this avoidance is due to the fact that some preservatives have toxins that can seep into the soil. The soil holds the plant's root system, which ultimately could send these toxins into the fruits or vegetables produced by the plants.

A variety of other materials may be used for the raised bed frames. Brick, pavers, cement block, and even two-by-fours are viable frame materials. Commercially, framing kits may be purchased at garden centers, home stores, and various other venues, including mail order and the internet. Along with the kits, there are new versions of raised bed garden kits and materials coming out all the time. So the options available for those who opt for the raised bed garden are great.

When building the raised bed frame, it is actually like building a box with no top or bottom. However, it should always have four walls, even if the bed frame is being built in front of a wall, as having soil piled against the wall itself could eventually lead to the wall rotting out or cause other problems along the way.

One thing to consider during construction is whether or not the frame will be **sunk into the ground** or **set at ground level.** Although this is basically a matter of personal preference, there are pros and cons in the decision to "sink or not to sink."

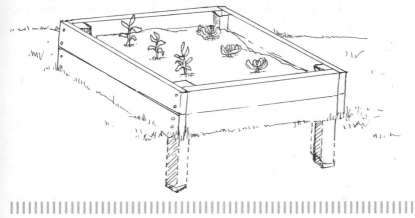

Setting up a raised bed garden. Illustration by Ariel Delacroix Dax.

If there will be problems with burrowing animals, then the best option is to sink the frame. To do this, the perimeter of the space should be dug down at least one foot so the base layer of the bed frame is below ground level. This will prevent animals from digging or burrowing under the bed and gaining access to the garden. From this point, the bed frame will be built as usual. And don't forget to allow for the sunken part of the frame in the measurements. For example, if a four-foot-high bed is desired, then the height of the entire sunken bed should be five feet. One foot below ground level and the rest above, which would leave you with the desired four-foot-high raised bed.

If burrowing animals are not a threat to the garden or sinking the bed would just prove too difficult a task, then the frame may be built totally above ground. In this case, the base of the bed would sit flat on the ground and built up from there. It is that simple. It is indeed less work to build the frame totally above ground, and neither way will have any affect as to how well the plants will actually grow. In the end, the best way to decide which way to erect the raised bed frame is to analyze the potential threat to the garden, and then choose accordingly.

Once the frames are built and secured, the next step will be to add the **soil mixture.** The ground should be loosened up at least a foot deep before adding the new soil mix, and make sure the depth of the soil added is adequate for the plant's root system.

A common mixture that can be made by the home gardener is as follows:

- Half organic matter (usually compost)
- Half soil (purchased or dug up)
- A bit of sand for drainage

There are a number of other "recipes" for soil mixtures available that can be found online, through the local extension office or by asking local nurseries, but this is the most common one used and the easiest for a novice gardener to obtain the materials for and produce. After a while, once becoming more comfortable in working with the soils, some gardeners will end up creating their own mixtures and ratios. If time is an issue, however, a premixed commercial soil may be used in the bed. If you are creating large beds or multiple beds, you may try to purchase loose, bulk soils by the truckload, as it will usually result in a better price. These may be found in larger garden centers or commercial operations that deal in landscape or soils. It should be noted at this time that if the new garden is to be a totally organic garden, then the commercial mix selected must be a totally organic mix. Although sounding strange, as who would think that soil is anything but organic, nonorganic soils may have nonorganic additives or other additives that do not fall within the organic guidelines. Should you want a true organic garden that would be certified, commercial soil mixtures that have unapproved additives will not pass muster and certification will be denied. If you are seriously interested in organic certification, the USDA website at http://www.usda.gov as well as other state and local organics websites will present any rules and regulations to get the garden certified.

On a final note, raised beds are a perfect solution for **rooftop gardens**. However, you must, and this is very important, be sure that the roof that will house the garden will be able to handle the extra weight, keeping in mind that not only will there be soil weight to consider, but weight from the plants as they grow and fruit and weight from the wet soil due to watering and rain as well as from the snow and ice that will collect on the dormant garden in the winter on top of the soil that is already there (for those gardeners in harsh winter climates).

In the end, although the raised bed garden may be a bit more work in the beginning, it is something that does not have to be rebuilt year after year, and it will save time and energy with weeding chores and watering. All in all, a very reasonable return in the end.

Straw Bale Gardening

Straw bale gardening is another gardening method typically used in conjunction with either a traditional or raised bed garden and not by itself.

You can garden with straw by planting in the bale itself or packing straw in a form. The straw needs to be soaking wet and kept wet. There is a process to this type of gardening, much more than sticking a plant in a bale of straw, but it can be a fun way to

A typical straw bale garden set-up. Photo by Judy Thomas (www.cvog.blogspot.com)

garden once you master it. Which method you use will depend on what you can grow. And the straw will usually need to be added to or replaced each season, again depending on the method used.

Container Garden

Although nothing new to the gardening scene, container gardening of fruits and vegetables has become a popular gardening alternative, especially for those gardeners with little or no space for planting. This method allows those with only a balcony space to raise vegetables or even fruits. Container gardening is also a good alternative for planting perennials such as mints, which are a garden favorite, but also tend to take over a garden space when planted directly into the ground and left to their own tendencies. It is also a good choice for those plants and fruit trees that cannot tolerate extreme cold weather and must be brought indoors or to the greenhouse during those cold months, as it makes the plants "portable."

If you decide you need to turn to the container gardening method for your growing needs, choices in what to grow are not

Cauliflower growing in a container garden. Photo by LollyKnit under the Creative Commons Attribution License 2.0.

quite as limited as you may think they would be. Because of the newfound popularity this method is finding, many more seeds and plants are coming to market labeled as suitable for containers. And even those that are not labeled as such can still work. In fact, this is a great time to experiment if there is space, time, and desire.

Many fruits and vegetables are conducive to being grown in a container, including tomatoes, peppers, a variety of berries, dwarf fruit trees, herbs, and lettuce, just to name a very few examples. Climbers and those plants that need support, like tomatoes or cucumbers, can have small trellises, stakes, or cages placed within the pots to support their growth in an upward fashion, just as they would grow in a traditional garden.

When planting in containers, it is important to be sure that the chosen pot will be adequate for the plant and root system going into it. For example, a tomato plant will need to go into a larger and deeper pot than an herb would, but in a smaller pot than a dwarf fruit tree or berry bush would need.

Here are some examples of soil depth for container planting:

- Lettuce and peppers: eight to twelve inches
- Beans: sixteen to eighteen inches
- Carrots: nine to eighteen inches (depending on type)

In container gardening, there may be one plant per container or there may be several. It all depends on what is being planted. In some cases, especially with herbs, edible flowers, or strawberries, multiple plants can be together in large planters. Some vegetables may be planted together as well if the pots are large enough. Some pots, such as **strawberry pots,** are specially designed to grow multiple plants within it. Although called strawberry pots, as that was what these pots were originally designed to grow, they will also work wonderfully as herb planters and would be a decorative addition to any porch or patio. However, whether holding one plant or multiples, the bigger and deeper the pot (in proportion to the plant's needs).

And to make watering even easier in pots such as strawberry pots, before adding soil, stand a small piece of PVC in the center of the pot. The piece should have holes drilled up and down, be about an inch in diameter, and should rest on the bottom of the pot and come one to one and a half inches above the top. Then fill with soil and plant as usual. When it comes time to water, simply pour water into the pipe, and it will evenly disperse throughout the pot. A great solution for difficult container-watering situations.

PVC watering system. Illustration by Ariel Delacroix Dax.

When the appropriate pots or planters have been selected for the plants and seeds, before planting, the pots should be examined to make sure there are drainage holes on the bottom (although sometimes there are drainage holes on the sides of the pots or on the bottom). These holes are necessary so water does not accumulate in the pot and rot out the root system, which in the end will kill the plant.

Most pots will already have some type of drainage; however, if not, there are a few things that can be done to remedy this. If the pot is made of plastic, holes may be drilled into the bottom.

If the pot is clay or ceramic, it may be possible to carefully drill holes in the bottom, but there would be a very good chance of the pot cracking and breaking. Instead, lay a few pieces of broken pottery shards, a few stones, or even some packing peanuts in the bottom of the pot before the soil has been added (if using packing peanuts, make it about one inch deep.) This will give excess water a place to go away from the soil and roots. In fact, even if there are holes in the pots already, it is still a good idea to place a few broken pottery shards over the holes. This will help keep the soil from coming through during watering or even when filling the pot. It should be noted that gravel should not be used, as it will not allow proper drainage.

Soil for the container garden can be the same mix as for the raised bed. If very large pots will be used (as for dwarf trees and berry bushes), a lighter soil mix would be recommended in order to allow easier movement of the pot or pots. (If the pot is quite large, it may be a good idea to put wheels under it as well.) The mix to be used may depend on the type of tree or bush, so it is recommended to ask a local nursery or extension office about

Hanging Buckets

In the last few years, hanging pots have become available, which allow you to grow vegetables (and some fruits) upside down, hanging from the bottom of the planter into midair. They are definitely space savers, but they can be quite pricey, especially if you want more than one or two of the containers. However, if you are on a tight budget, you can build them yourself from found objects around the house. A five-

Plants growing in hanging buckets.
Illustration by Ariel Delacroix Dax.

gallon bucket (with handle intact) works great. You should be sure that the bucket is clean and was never used for anything toxic or poisonous, as it could affect the soil and the plant (and, if it goes through the root system, can harm those who eat the food produced in the container). Drill a hole in the middle of the bottom of the bucket, about one and a half to two inches in diameter. If you are afraid of drowning your plant when you water, you may also drill a few small holes in the bottom for extra drainage. Cover the bottom on the inside with a piece of garden cloth cut to fit. Insert plant in the hole, cover with lightweight soil, and hang.

the tree or bush in question. Again, either a commercial mix or homemade mix may be used if the correct ratio is known.

At this point, the pots are ready to fill with the plants of choice, which will create the container garden. It should also be mentioned that while container gardens provide those with little or no space a way to still garden and is an excellent solution for supplementing the family food supply, it is unlikely to provide all

|||||||||||||||||||||||||||||||||||||||||||||||||||||||||||||||||||||||||||||||||||||||||||

Hydroponics and Aquaponics

One container garden alternative is hydroponics. Hydroponics centers around growing plants without soil, usually indoors with an elaborate setup. Although there are more commercial greenhouse growers using this method, due to the cost involved in the setup of hydroponics, few backyard gardeners use this method, although small bits and pieces of hydroponics may find their way into the home garden. With hydroponics, there are a number of special needs and supplies, and although it can allow for year-round food growing, many gardeners find this method too expensive or impractical to use as an entire or even partial home garden.

Aquaponics, however, is making appearances in some backyard gardens and even in homes. Aquaponics, as with hydroponics, is a method of soilless gardening in containers. But unlike hydroponics, this method also includes raising fish in conjunction with the produce within the same system, using the water from the fish to feed the plants, which in turn filters the water that is then returned to the fish as clean water.

Through aquaponics, two types of crops are raised: vegetables and/ or fruits and fish. And although commercial setups can be expensive, there are all sorts of working examples of home-built systems in all shapes and sizes, with some even using recycled or repurposed materials.

And although many larger setups use tilapia as their fish of choice (as the point of aquaponics is two food-production systems in one area, working off of each other), if you are setting up just a small system and don't much care about having fish to eat, you can use larger goldfish instead.

For anyone interested in learning more about hydroponics and aquaponics, there are many resources on the Web for gardeners of all experience levels and all budget sizes.

|||||||||||||||||||||||||||||||||||||||||||||||||||||||||||||||||||||||||||||||||||||||||||

the food a family will need or supply as much as other gardening methods. However, whatever the container garden can supply will be worth the effort.

Vertical Garden

Just as the name states, these are gardens that grow "up," using lattice, fencing, stakes, or anything else that may be safely used to fully support the plants and their fruits as they grow. **Vertical gardening** can work with traditional gardens, raised beds, and even container gardening.

So what exactly does vertical gardening entail? Basically, the garden is planted as normal, usually in the traditional or raised bed style. However, as the plants begin to grow, they are trained to grow upright or vertically. This is achieved through various supports placed behind the plant. As the plant grows, the limbs are braced by and usually fastened to the support in an upright position.

This method not only works with those plants that will naturally grow upward, like tomatoes and peppers, but for vine plants that usually grow along the ground, such as cucumbers, squash, and melons. Not only will growing up give more room in the garden and prevent you from stepping on fruits and vegetables normally laying on the ground, but it can also increase the yield of a plant and make harvest time much easier, especially for those gardeners with physical limitations or disabilities.

When searching for supports for the vertical garden, the materials and types may vary and even become a bit creative. Whatever the support in use, it must be strong enough to hold the weight of the mature plant and its fruits, both dry and wet (remember, plants after a rain will be heavier).

There are some supports made commercially, such as tomato cages. Not only will they do a good job of supporting tomato plants, but the cages will also work well with other plants that

Tomato cages. Photo courtesy of www.tomatoguy.com

would need support, such as peppers and cucumbers. Other supports may be created from things that you may already have. Posts and string make a great system for peas and beans.

When using stakes in the garden, there are a variety of choices. They can be wood, metal, bamboo, or even broken branches. (If using branches, they should be green for optimal strength. Starting with old, dry branches could result in the branch cracking under the weight of the plant.)

Any stakes used need to be taller than the mature plant will be, simply due to the fact that if the stake is shorter than the plant, support will be lost. However, sometimes a plant will grow larger than what is normal and the support may be short. This happens and there is really nothing that can be foreseen in this situation.

When placing stakes or tomato cages, make sure that they are sunk sufficiently into the soil, be it in the garden or in a container. Otherwise, as the plant grows and gets heavier and heavier, the support may pull out of the ground, causing the plants to fall and even entangle with each other. Should this actually happen, it isn't a disaster, and if the plants are too entangled, it may be best to just leave them alone. It could make harvesting a bit tricky, but it is better than broken plants.

If stakes are being used for peas or beans, you will need to do more than just stick some wood into the ground. These plants need to attach themselves to something to properly grow. **Teepee structures** may be built from bamboo or branches, allowing the plants to attach and grow up. Stakes may also be fastened together with another stake running along the top and securely fastened. Then garden twine should run from the top of the frame to the ground, fastened with grounded clothespins or small tent stakes. The string will allow the runners a place to go and grow, usually finding their own way grabbing the strings. But sometimes they may need a bit of direction. Should this happen, carefully wrap or fasten the runner to the stake or twine. From there the plant should take off on its own.

Lattice work is nice and sturdy for heavier needs, such as squash, melons, and eggplant, among others. However, the heavier

Posts and string supports for pea plants. Photo courtesy of Jennifer Cooper/Pumpkin House Studio (www.pumpkinhousestudio.com)

weight of lattice needs to be used, making sure it is untreated. Heavy netting or chicken wire may be attached to a fence or wall. Garden arches can also be a great base support for plants, especially perennials like berries and grapes. Trees may also be trained to form over the arch, creating not only a functional food support, but a beautiful architectural piece for the garden or yard. Even other plants, such as low evergreen bushes, may be used as support.

If using a **trellis**, make sure that it is tightly fastened to a wall or fence or otherwise securely standing in the garden. As there are also differences in the thickness, weight, and materials of trellises, specific cases must be taken into consideration when making the selection. For example, there is a great difference in weight of a squash plant trained on a trellis and the weight of peas. You will have to decide what will be growing on the trellis, then choose accordingly. It is also worth mentioning that making a simple frame for the trellis will add strength to the piece, no matter the size. However, the bigger the trellis, the better it would be to have a frame.

One thing to keep in mind, however, is that some plants will also need support for their fruits as well. A good example is melons, whose weight could cause a slight problem when hanging vertically. This can easily be remedied by creating slings for the

Espalier Gardening

Finally, for anyone who wants an orchard, is limited in space, but does not want to grow trees in containers, espalier may be an option to consider. Although it is rather time consuming to start out, espalier trees are trimmed and trained to grow up against a wall, fence, or trellises, while still producing plenty of food for you and your family.

Most any fruit tree may be trained to the espalier method, as there are many different ways to train trees correctly. Again, for anyone who has an interest, there are any number of websites as well as books and some magazine articles that will discuss the espalier method in more depth and detail.

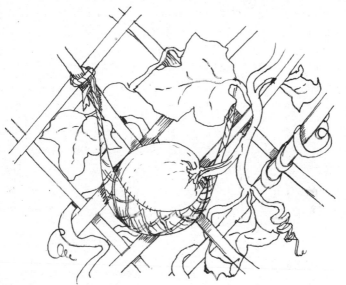

Sling for heavier fruits and vegetables. Illustration by Ariel Delacroix Dax.

fruit, using wide-hole netting that will allow rain to easily pass through. (Some of these methods will be further covered later.)

In wrapping up, vertical gardening is probably the best way to get more out of less space, up the yield of whatever is to be grown, and make the harvest much easier to pick than would be the case for a traditional or raised bed garden. However, it may be used in a limited basis for container gardening as well.

CHAPTER 4

DECIDING WHAT
TO PLANT

Now that you've decided on your desired gardening method (or methods), it is time to determine what will be in your garden and where. You have already put together your "wish list" in Chapter 2, but now is the time when reality needs to set in as to what will actually work for you. As a new gardener, you may be a bit unsure about whether you can devote the proper time to your garden. A good rule of thumb for beginners is to start small and stay with basic, easy-to-grow vegetables and herbs. Tomatoes, cucumbers, leaf lettuce, and many herbs are easy to find in both seed and plant form, being easy to grow either way you choose. Otherwise, there are hundreds, if not thousands, of varieties of vegetables, herbs, and fruits, as well as edible flowers, that you can select for the garden.

Types of Plants

When selecting plants and/or seeds for the garden, it is important to remember that there are four types of plants. They are:

- Annuals
- Biannuals
- Perennials
- Tender perennials

Annuals are plants that come up for one year or season only. The following year, new plants must be replanted as the plants from the season before would have died off after the first hard frost or during the winter. Most vegetables will fall under this category, especially when growing in a region with four seasons. (In milder areas with three seasons or less, annuals may last longer than a single season before replacement is required.)

Biannuals are plants that will go to seed in the second year. Foxglove, a medicinal plant that is also commonly grown for garden decoration, falls into this category, as do carrots, onions, and beets.

Perennials are plants that will come back year to year. Many herbs and some edible flowers fall under this category, as well as vegetables such as asparagus and lovage. Many berries also fall into this category, as do all fruit trees.

Tender perennials are perennial plants that are grown as perennials in mild climates and annuals in cold climates, as they cannot survive extreme cold and snow. Some excellent examples of tender perennials include the rosemary plant, French tarragon, and various lavender plants.

There are also two more types of plants to take into consideration. They are heirlooms and hybrids. These plants may be annuals, biannuals, perennials, or tender perennials. The differences in these plant types are as follows:

Heirloom plants are pure, meaning there has been no crossbreeding with other plants. Heirlooms can date back decades or centuries. They may have unusual names as well as unusual colors or markings. Heirlooms are making a comeback into home gardens. They can be heartier to weather conditions and tastier than hybrids. However, they can also be a little smaller in fruit size and have a few problems in extreme situations. Heirloom seeds may be saved for planting the following year. They may also have some very colorful or interesting origins.

Hybrids are crossbreeds between two types of the same

species. Hybrids usually date back decades, although some may go back farther. They may not be as interesting to grow as heirlooms, but some have been bred to stand weather extremes. Flavors may vary, and many gardeners who have grown heirlooms think that many hybrids have lesser flavor. Hybrid seeds may be saved, however, when planted, and results may be disappointing, as there may be little or no development.

As a result, the saving of hybrid seeds is often unpredictable. If you have the time and some extra space, it could be fun to experiment and see what happens. But keep in mind that results will be mixed, maybe even nonexistent. And if some seeds do work with one planting, it doesn't mean the same results will be obtained the next.

As a result, the use of heirlooms or hybrids is up to your personal preference. Some gardeners will decide to grow only heirlooms, while some may want only hybrids. Many will opt for a combination.

Tips on Choosing Your Plants

When planting a garden, the plant category usually does not come into play unless you plan on planting a total heirloom garden. However, when planting herbs and edible flowers, whether a plant is an annual or perennial may be something to be taken into consideration, especially if you do not want to have to replant your herb garden year after year.

If you are an inexperienced gardener, you may want to look at easy-to-grow vegetables for your first year to get your feet wet.

The following are some examples of fruits and vegetables that are easy to grow and almost always produce a successful harvest:

- Most herbs
- Lettuce
- Carrots
- Zucchini
- Beans
- Tomatoes
- Peppers
- Spinach
- Strawberries
- Blueberries
- Raspberries
- Blackberries

Seeds, Plugs, or Plants?

After you have decided what you will be growing, the next step is to decide whether to start with seeds, plugs, or plants. And it really isn't as complicated or intimidating as it sounds.

First, what do each of these terms mean? **Seed** is pretty much self-explanatory. Most people,

A handful of seeds before planting. Photo by Farmer Dave (www.family-gardens.com)

gardeners or not, are familiar with seeds. They are the base on which all plants start. Seeds are an inexpensive way to start, and you can have quite a variety of produce in your garden starting this way.

Plugs are very young plants, also called seedlings, due to the fact that they have just sprouted from the seed. Plugs will be small and more expensive than seeds but cheaper than plants. Plugs may need a bit more babying than plants in the garden to start, but will usually fare quite well and will give you a slight head start over starting with seeds. Plugs are

Plugs awaiting transplant. Photo by Talitha Purdy

sometimes sold as singles, but may also be found in four- and six-pack containers. Flower plugs may already be in some bloom at this stage.

Plants are larger than plugs and usually sold singularly in small pots. Some may even be large enough to already have a number of blossoms growing, and may even have little vegetables or fruits developing. Plants are the most expensive way to go as the grower has had to take more care and time with them, not moving them to market as quickly as plugs.

For the most variety, seeds are definitely the way to go. Seed catalogs have an enormous selection of all different types of vegetable and herb seed. For the experienced gardener or the new gardener who isn't afraid to jump right in, seeds are the best way to get the most variety in your garden for the least money. Seeds may be started indoors early in the season in seed trays (in fact, depending on climate, this is a must for some) or may be planted directly in the ground. (More details on starting from seed can be found in Chapter 5.) Either way, the package will indicate the best time and method for starting the seeds, usually having some type of growth chart on the back.

For quick-growing plants, planting directly into warm soil is fine. Some seeds can even be started in cooler weather, especially greens like lettuce and spinach. However, for those seeds that need a longer growing time that you may not have due to climate, the seeds would need to be started earlier indoors, in the house, or greenhouse.

Although seed will give the greater selection of prospects, you may feel a bit intimidated starting everything from seed, particularly if you are a first-time gardener. And if the seeds don't germinate, which can happen, especially in bad years, you may feel defeated. But then again, there is no greater sense of accomplishment than taking a plant from seed to harvest.

So what to do? Try using a combination. Choose the fast-growing plants that can be sown directly in the ground to start from seed. The more difficult plants or those whose seeds need to be started indoors due to time restraints can be purchased as plugs or plants, which can then be placed directly in the garden.

By using the combination of seeds, plugs, and/or plants, you can still have the experience of growing from seed without having to start everything from seed.

Does this mean all first-time gardeners should not start their entire garden from seed? Of course not. But for those who are a bit nervous starting with seed or are afraid of little success their first time out, but still want the seed experience, combining methods really is the best of both worlds. Then once you have your first garden under your belt, the following year you will feel more comfortable and confident with your garden. You may also feel confident enough to grow even more from seed. (And are those seed catalogs fun!) Once you start, you will wonder why you waited so long to grow your own plants from seed.

Understanding the Hardiness Zones

Before making the final decision on what will or will not be part of the garden, you should review the USDA Hardiness Zone Map. This map shows the United States broken down into regional colors and temperature or hardiness zones. All plants have a **"zone hardiness,"** meaning the best areas for a particular plant to grow based on climate and temperature. Using this map and the zone-hardiness numbers that can be found on many seed packets, plant packs, and singles, you can determine if a particular plant can be grown in your garden.

It should also be noted that some plants which may not be outdoor zone hardy for a certain area may still work as a house or potted plant. A good example of this is keeping a dwarf citrus tree in the north. It will do fine outdoors in the summer, but will not survive the first major freeze. However, many northern gardeners still keep citrus by keeping the trees potted, allowing the gardener to bring them outdoors in the good weather, then return them indoors (to either the house or greenhouse) before the first freeze or as the temperature begins to drop dramatically.

The hardiness zone map is available as a free download and may be printed out. It may be found at the USDA website at http://www.usda.gov.

Companion Planting

Companion planting is the age-old method of planting plants together that will "get along" and can be beneficial to each other during their time in the garden. It also takes into consideration those plants that would not get along due to, for example, competition for the same nutrients. That being said, companion planting isn't necessary in the garden unless there are plants that are truly incompatible. However, you will find the extra time and effort put into this arrangement to be worthwhile.

Companion planting may be used with any of the four gardening methods, and if done properly, can help to make your work a little easier, especially when the companion plantings aid in pest control.

One plant that works well as a general companion with any plant is the marigold. Marigolds not only repel pest insects, such as aphids, but attract beneficial insects as well. (And marigolds are edible too!) In a traditional or raised bed garden, scatter them among the other plants. With vertical gardens, they can also be scattered among the plants. When container gardening, if containers are sitting on the ground, simply plant the marigolds around the pots. If the containers are sitting on decks, porches, or some other way off the ground, intermingle pots of marigolds with the vegetable containers. Or they can be planted directly in the pots with the vegetables or herbs.

Other examples of plant companions that work well together include:

- **Tomatoes and onions:** Onions help keep slugs off tomato plants.

- **Beans and corn:** When given a head start in growing, corn will become the support system for the beans, with the bean plant growing up the corn stalk. These are not only good companions, but also a great space saver in a small garden.
- **Cabbage (and cabbage family):** Plant aromatics such as sage, thyme, or lavender to keep cabbage worms away.
- **Garlic and roses:** Garlic keeps the Japanese beetle away from the rose bushes.
- **Peas and squash:** Trellis well together in vertical gardens.
- **"The three sisters":** Corn, beans, and squash with beans using corn as a support and squash planted around the bottom.
- **Basil and tomato:** When planted together, basil is said to improve the flavor of the tomato.

This is just a very minute sample of companions. There are a number of lists and charts online, giving hundreds of combinations of plants that will work well together. You can also contact your local extension office or nursery for further information.

Just as the previous list shows plants that are compatible with each other, there are also plants that should be nowhere near one another.

Companion plants help in keeping gardens healthy. Illustration by Ariel Delacroix Dax.

Some examples of plants that are not compatible include:
- Strawberries and cabbage
- Potatoes and tomatoes (common fungus)
- Cucumbers and sage (inhibits cucumber growth)
- Onions and beans (inhibits bean growth)

CHAPTER 5

LET'S CREATE A GARDEN!

········· · · · · · · · · · · · ·········

With the layout of the plants figured out, it is now time to create and plant your garden. When deciding on the location of your garden, keep in mind that the plants should receive at least six hours of sunlight each day, if at all possible.

Prepping the Soil

Another important consideration for the location of your garden is the type of soil that the new garden plot holds. This will not be much of an issue with container gardens or raised beds, as those will have newly mixed soil added to the frame or pot, giving you full control of the soil the plants will be in. However, with the traditional garden, you will not have this luxury and will be working with what soil is there, making any necessary additions if and when necessary.

Ask yourself whether the soil is sandy or clay. Does the pH indicate a problem with acidity or alkalinity? Do nutrients need to be restored? Getting the soil back to optimal condition can be as simple as adding a bit of fertilizer, or much more complicated. It will all depend on soil type, what is being planted, and even location, as well as how worn the soil really is.

If there is any question as to how fit the soil is for planting, although there is no shortage of information online and in books and magazines, beginner gardeners will most likely find it much easier to have a local extension office or greenhouse assist with the situation. They can guide you in the appropriate direction as to testing the soil (kits are available at home stores, greenhouses, and even through some extension offices) and identifying any nutrient deficiencies or any corrections that the soil may need for the specific location in order to obtain the most from your garden plot.

Planting Your Seeds

Planting seeds using a dibble.

No specific tools are necessary to plant seeds. As each seed has specific requirements, such as the depths at which they must be planted, you should read each and every package carefully before planting your seeds. If you need a hole for your seed, a small spoon (or even your finger) works well. You can also use a **dibble,** which is a simple tool that pokes a small hole in the ground to hold the seed. But unless you already have one, it is not necessary to go out and buy one. Just make a hole at the correct depth (according to the seed package), drop in the seed, cover, and water.

Testing Seeds

Not sure if some older seeds are viable? Test them before planting.

Needed:

- Quart storage or freezer bag, zipped type
- Paper towel, one sheet
- Four to six of the seeds in question

Run the paper towel under water until well moistened. Towel should be thoroughly wet, but not dripping. If this does happen, lightly ring out, but not so much that the towel dries out.

Lay wet towel on counter or tabletop. On one half, lay the seeds out, leaving some space between each. Then fold the other half of the towel over the seeds and allow to lie directly on them.

Carefully lay the towel with the seeds inside the bag and seal. Do not allow to dry out. If towel begins to dry during testing, add a little water to the bag to moisten again. But not too much. Towel should again be quite wet, but water should not pool in the bag.

If the seeds are still viable, they should sprout inside the bag within their normal germination time (give or take a few days). If nothing happens, then the seeds are obviously not good and should not be used. If they do sprout, you have viable seeds. And don't throw those test sprouts away. Instead, carefully remove them from the towel and place them in small, soil-filled seed-starting pots (with potting soil, one in each pot). Let them grow until they are large enough to be transplanted into the garden or into a larger pot, if using container gardening. Remember that these plants will have a head start on the others, so there may be a bit earlier fruiting on the test-seed plants.

When testing seeds with this method, check the sprouting time on the package. Make sure that enough time is allowed to start additional seeds for the gardens should the test seeds be viable. Do not test seeds during the time frame that the seeds should be started for the regular planting time. Ideally, seeds should be tested at least two weeks before the gardener plans on starting the garden seeds.

Seed-Starting Tips

When starting seeds, it is not necessary to purchase starter pots. Household items can be recycled and repurposed for use.

Egg cartons, foam, paper, or plastic, make great seed starters. Remove the top of the carton and set aside. Poke

Starting your seeds using egg cartons. Photo by Cat Brimhall, Set the Trail (www. setthetrail.blogspot.com)

a small hole in the bottom of each egg compartment. Fill with potting soil or seed-starting soil, then drop one seed into each compartment. Use the carton lid that was set aside for the drip tray, placing it underneath the seed-filled carton. If using paper cartons, line the lid with foil or plastic wrap to prevent leaking.

It cannot be stressed enough how important it is to allow the hardening off of seedlings before planting, especially when they have spent their first few weeks in a warm building or have been kept indoors all winter long.

Hardening off a plant simply entails getting the indoor plant gradually used to the cooler outdoors before either planting the seedling or leaving the containers outdoors for the season. Although the use of cold frames were mentioned earlier, if one is not available, then the gardener needs to take the plants outdoors on a daily basis, bringing them back in during the still cold nights, as frost threats may still linger in some regions.

After the threat of frost is over, and the nights begin to get warmer, the plants may begin to remain outdoors or, if still too cool to stay outside, be put into a garage or other outbuilding at night, preparing to be transplanted into either the garden or a container.

When planting, seedlings need to be removed with their little

II

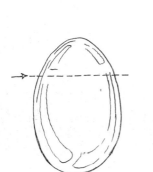 

IIIIIIIIIIIIIIIIIIIIIIIIIIIIIIIIIIIIIIIIIIIIIIIIIIIIIIIIIIIIIIIIIIIII

Egg shell seed starting. Illustration by Ariel Delacroix Dax.

dirt ball from the foam carton. Paper cartons, however, may be cut apart and planted as a little unit. However, decomposition time of the paper may vary. It may help to make little cuts in the paper before planting.

While on the subject of eggs, egg shells may also be used to start seedlings, using the egg carton as the holding tray. The trick to using the egg shell so there is plenty of germination depth is to break the egg closer to the top rather than midsection.

As the egg shells are saved, they should be carefully washed to remove any residue from the inside and left to dry. Once dry, put potting soil or seed medium in the egg shell (remember to poke the small hole in the bottom of the shell first) with a seed. Handle as any other seed-starting method. When seedling is large enough to transplant (and the outdoor temperature is also correct or the gardener has a cold frame to harden off the plants), there is no need to remove the seedling from the shell, as the entire piece may be planted in the garden or container. The calcium from the shell is good for the plant.

The small four- or six-pack seedlings may also be reused for seed starting. That is if the frail packs survive their first use.

However, before reusing, clean in a very light bleach and water solution or use an organic disinfectant that would be safe to use with plants, and let dry. This should kill any unwanted bacteria or other potential problems that may have remained with the pots.

The domed plastic cups that many frappes come in make terrific mini-greenhouse seed starters. However, because they do take up space and you would need quite a few to start all of your seeds, these cups are best for those times when only a few seedlings of a certain plant are wanted. For example, you may only want two or three pumpkin plants in the garden, especially if it is a small garden space. Using these cups is a great way to start up those seeds (and pumpkin seeds do start well in these cups).

Problems to Watch for When Transplanting

Sometimes if a plant has been in its little starter pot for too long or has had too much light before transplanting, it may grow tall and spindly, especially tomatoes and peppers, but this may also happen with eggplant and other vertically growing plants. In order to give the plant a little more sturdiness and strength, they should be planted a bit deeper than just root level. The plants will grow stronger and loose the long, spindle look, filling out nicely as they continue to grow.

Propagation

Another way to start new plants is through propagation. **Propagation,** which may also be referred to as rooting, is taking a piece of a plant to create a new plant. Basically, this process involves taking either root or leaf cuttings, depending on the plant, placing in potting soil, keeping it moist, and letting a new root system develop. The root or cutting may also be placed in a cup of water for rooting. This method can be used with herbs (especially mints, either leaf or root), tomatoes (leaf or small stem piece), pineapple (top), and even some edible flowers. As this is only a small sampling of plants for propagating, if the gardener is curious about a plant, it can easily be researched online or inquired about at a local greenhouse.

Propagation can be lots of fun and yet another way to share plants with your fellow gardeners. All gardeners should give it a try, if for nothing more than the experience.

Gardening Tools and Equipment

Using a tiller on a traditional garden. Photo by Mantis (Schiller Grounds Care, Inc.)

Whether your garden is large or small, in the ground or in pots, you will need at least a few pieces of equipment. The equipment needed will obviously depend on the type and size of your garden (and in some cases, your budget). The equipment doesn't have to be expensive, although many times, cheaper equipment will not last, invariably breaking in the midst of a major project. Some necessary equipment may not even need to be purchased at all.

For example, most who put in a traditional garden will till or turn the ground. The ground may be turned by hand using a shovel, but it is hard work and can require a great deal of time, so many gardeners will opt to use a tiller to break ground.

Tillers come in assorted sizes and price ranges, some being a bit pricey, especially for the gardener on a limited budget. For most gardeners, the tiller is used only a few times a year at best, so you may choose not to make the investment into something that will get such a very small amount of use.

If you do decide to purchase your own tiller, they can be found at farm and garden stores, farm-equipment stores, and even some home-improvement stores. Some may even be purchased direct from the manufacturers. If you decide you do not want to purchase your own tiller, there are other options available to you.

One option is to rent a tiller. Most equipment rental centers will have tillers available for rent and sometimes in various sizes, allowing you to rent the best machine for your needs. Rental is usually by the day, but depending on the size of the area to be turned, soil conditions, unforeseen problems, and cooperation from weather, this job can most likely be completed within a single day.

Another option is to hire the job out to someone with the proper equipment. There are a number of tiller owners, most of them gardeners themselves, who will provide tilling services. They will usually first make an appointment to come to your home to assess the area and provide a quote. Prices will vary. Some will charge by the hour, while others will charge by the job. But in order to make a quote, the person for hire will need to know the specifics of your garden, including having the size and shape of your garden already marked out so the person giving the quote may do so accurately.

People who provide tilling services are not too difficult to find. If you can't get a recommendation from a friend, check feed- or garden-store bulletin boards, local newspaper ads, extension offices, garden centers, nurseries, and even front yards, as many who offer tilling services will have signs posted in their front yards. Another place to look would be landscaping companies, which may offer tilling as one of their regular services or may even be willing to take the job on a weekend, if it is not a regular offering.

Other than a tiller, what other tools are needed in the gardener's arsenal? Below is a brief list of some of the basics needed for any gardener's cupboard. There are, of course, many more garden "toys" available, and actual needs will depend on the type of your garden.

An assortment of common gardening tools. Photo by Melvin Baker ("Barefoot In Florida") under the Creative Commons Attribution License 2.0.

Here is a list of some of the most common gardening tools:
- Shovels, spade, and hoe
- Garden rake, lawn/leaf rake
- Weed popper (although not a necessity, this can make some weeding easier)
- Pots of various sizes, shapes, and materials
- Peat pots and peat discs, other seed-starting material
- Hose with changeable spray head
- String, stakes, and tomato cages
- Knee pads (highly recommended if there is discomfort kneeling on the ground or if the ground is hard)
- Garden clippers
- Garden markers and labels (it is a good idea to label rows or pots of plants and seed as they are planted. The empty seed packets may be used as a marker, as can ice-cream or craft sticks, the tags that came with the plants, writable plastic or ceramic markers, or even the fancy decorative markers that may be purchased.)
- Buckets and baskets
- Wheelbarrows or other carts

CHAPTER 6

MAINTAING YOUR GARDEN

. . . .

A t this time, the soil is ready and your plants and seeds are in. It's now time for Mother Nature to do her thing...with a little help from you handling the standard chores of watering and weeding.

Watering

The amount of watering necessary for a garden will depend on the type of garden, what is growing, and, of course, the weather. No matter what, a new garden will need lots of water so the seeds can germinate and the new plants can establish themselves. While you can stand outside, hose in hand to water (which sometimes may be the only option), there are some easier ways to get the job done on traditional, raised bed, and vertical gardens. Two of the easiest and least expensive options are the soaker hose and sprinklers.

Soaker hoses, for the most part, look like a regular garden hose (sometimes with a different texture) and attach to the faucet just like a regular garden hose. Many times the hose will be black to blend into the surroundings. The difference is that the soaker hose is porous. This porous hose lies in the garden on top of the soil. When the water is turned on, it seeps out of the hose slowly and gently, soaking the ground around the plants. When the ground is saturated, just turn the water off. The hose may stay in place until the next time or can be easily moved to a new location that needs attention. Many gardeners will leave these hoses laying permanently in the gardens whenever possible, making the job much easier and even less time consuming.

If you're on a tight budget, an old, leaky garden hose may be made into a soaker hose, performing the same duties as a purchased one, while practicing a little recycling as well. Just make small piercings down the hose (a very sharp object will be needed for this, so use caution), making sure that the piercings are small

A typical soaker hose. Photo courtesy of Porous Pipe Ltd.

enough so the water gently trickles, but does not gush, while in operation. Pierce all the way to the end of the hose, and you have a homemade soaker hose for the garden. Simple, money saving, and earth friendly too!

Sprinklers can also be a great time-saving option. However, it should be noted that some plants, like tomatoes, prefer to be watered from below. The sprinkler does not have to be an elaborate system and small, noncommercial yard and garden types are available at very affordable prices for home use. These sprinklers come in various sizes and spray patterns, with some sprayers being stationary, some swinging back and forth, and others oscillating. These sprinklers simply attach to the garden hose. You can place the sprinkler where needed and simply turn on the water source. As with the soaker hose, the sprinkler may be moved as necessary. It is again important to remember that some plants prefer to be watered from the bottom at the soil level, and watering from the top could cause fungus to grow on the leaves of those plants. If you are unsure about the correct watering style of your plants, information can be found online, from garden centers, or through extension offices.

Watering container gardens is different from watering the other garden types. As mentioned earlier, good drainage is a must and can mean the difference between a successful planting and a plant that dies in the pot.

Container plants tend to dry out faster than the traditional, raised bed, or vertical methods, so they need closer tending than the other methods and should be watered more deeply (although some mulching will help keep moisture in). Always be sure to check whether the containers even need watering, as sometimes the surface may dry while the surface below is still quite wet. An easy way to check this is to simply take a pencil or slim dowel and stick it down into the soil. If it comes up wet, no water is necessary for now. If it comes up muddy, that usually means too much water has been added, and the plant may actually need to dry out a bit so the roots do not rot. If it comes up dry, it is time to water.

Watering for container gardens may be done with a hose, a watering can (if there are only a few pots; otherwise, it could be too labor intensive), or a soaker hose rigged up along the tops of the pots. The PVC pipe method discussed in Chapter 3 is another easy and inexpensive watering method. If budget allows, self-watering pots may be purchased. If your budget doesn't allow this, there are many examples of self-watering pots online along with instructions on how to build and use them.

Another self-watering method, which would be particularly helpful if you are going away for a weekend or on a short vacation, is a simple, **water-filled sandwich bag**. Using either a zip-type or press-lock-type bag, fill the sandwich bag with water and poke a series of small holes into the bag (try using a tack along the bottom of the bag). Again, you do not want the water to be gushing, just trickling. The bag is then laid in the pot and left. Working somewhat like a soaker hose, the bag will slowly and gently release the water into the pot, trickling in and keeping the plant moist over a period of time.

Glass watering globes also work well, and will also slowly release water into the pots. Although these globes are fairly inexpensive, if there are many pots that will need attention, several globes will be needed. Usually one globe per pot is sufficient; however, very large pots may need at least two.

A selection of watering globes. Photos courtesy of PlantNannyInc.com

In general, when watering, plants should get between one to two inches of water from either rain or one of the watering methods mentioned. In extremely dry weather, more watering than usual may be necessary. If there has been an overabundance of rain, some plants may eventually need to be moved undercover so they do not get so wet that they end up with root rot. When watering is necessary, be sure it is done either in the early morning or late afternoon/early evening. If plants are watered in the heat of the day, they may end up burning in the sun if they have wet leaves. Many times this can lead to the loss of the plant. (Note that, when transplanting tomatoes to the garden or to the container, it is advisable to remove the bottom leaves from the plant. Not only will this direct the nutrients to the more desired upper leaves, it also prevents problems that may occur due to the bottom leaves holding dirt and water from the watering process. Although this is not a necessary step, it is one that should be remembered, as it will benefit the plant.)

Watering all boils down to this (with no pun intended): common sense prevails. Check the soil regularly to get an idea as to how much water the plants need and how often. You will soon see and learn exactly what your garden likes and doesn't like just by paying attention to how the plants and soil look. In just a matter of time, you will be able to tell the needs of your garden many times from a quick glance or by simply sticking a finger in the soil.

Weeding

Perhaps the one job in the garden that seems never-ending is that of **weeding.** Hot or cold, wet or dry, weeds will always appear, making it seem as though with each weed pulled yesterday, two more appear today. Of course, this isn't so, but particularly with the traditional method, weeds will thrive even when the vegetables do not. You will soon learn that the nasty little secret of weeds is that they are resilient little critters, and their defeat seems to be a never-ending battle. However, there is a way to help keep the infestation of weeds to a minimum without using chemicals. It is called mulching.

Mulching is done by placing a layer of something over the soil, such as wood chips, plastic, or newspaper, to help prevent weeds from appearing in the garden. This may not completely eliminate the weeds, but it will make a huge difference. However, it is a bit labor intensive when initially laying the mulch.

Perhaps the least expensive method of mulching, and a great way to practice recycling, is to lay newspaper (approximately five or six pages thick) on top of the soil around the plants. Once the newspaper is laid, a light layer of soil or other organic mulch should be laid on top, mainly to prevent the papers from blowing out of the garden during a wind. As the seeds, plugs, or plants are being put in, a hole needs to be put through the newspaper

Newspaper mulching. Photo courtesy of GardenPartner.com, www.gardenpartner.com

to allow planting of each piece. When planting seeds, make sure that the hole in the paper is big enough so, as the seeds germinate, they grow up through the paper and have plenty of space to do so. Of course, the layer of soil or mulch laid over the newspaper can also be laid over after the plants or seeds are put in, but this is up to you.

When using newspapers for mulch, the paper and the soil need to be thoroughly saturated with water before planting. As the watering is being done, it is critical that you make sure that the water is penetrating through the water and sufficiently soaks the soil. You may prefer to soak the soil first before laying the newspaper. This will allow certain saturation of the soil, and then the paper can be laid and saturated. Again, this is your choice. However, it is important to remember that each and every watering needs to saturate both the newspaper and the soil. Newspapers in the garden will last up to two years and need not be removed at the end of the season. Upon next planting, if more paper is necessary (as it will compost into the soil over time), just lay it over what is left from the previous year. In this case, five to six sheets may not be necessary the next time around.

Another method for mulching, albeit a bit more expensive than using newspaper, is the use of **garden cloth.** Just as with newspaper, garden cloth is laid down over the soil before the plants are put in (it is not quite as easy to lay the cloth after the plants are in). As with the newspaper, slits or holes should be cut into the cloth for planting.

The cloth may also be left lying on top of the soil or covered with a light layer of soil. Usually, the cloth is left on the soil, then held down with special pins made to secure the cloth to the soil. As the fabric does not get as heavy as the newspaper when soaked, it does need something to help keep it down. However, as the cloth may be a lot more porous that the newspaper, the water may soak through into the soil a bit easier.

The life of a garden cloth depends on the quality, placement, and how it will be used. The package the cloth comes in may give an idea of the lifespan of the cloth, but the average lifespan may be anywhere from a year to three years.

A third option for mulching is the use of **black plastic.** Although this does a great job keeping the weeds out and moisture in, black plastic does have a few drawbacks. While the black plastic will help maintain the soil's moisture, it also helps to keep in the heat. Therefore, it is important to make sure that any plants used with this method of mulching will be able to tolerate the warmth that the soil will retain. Tomatoes, melons, squash, strawberries, cabbage, and many herbs are examples of plants that will tolerate the heat generated by the plastic.

But just as the plastic will hold heat and moisture, it also will not allow moisture to penetrate the barrier, so neither watering nor rain will be able to do a good job in giving the plants their daily drink. As a result, some sort of irrigation system should be set up for use with this type of mulching. A simple solution is to run a soaker hose under the plastic, which will keep the plants hydrated. However, you must then keep close track of how wet the soil is and water accordingly, as it will not dry out as fast as other methods of mulching.

If plastic mulch is your method of choice, it is important that the plastic be black. Although clear plastic will hold heat and moisture, it will really do nothing to inhibit weed growth, allowing the little pests to grow right under the plastic. Why? Light can still penetrate clear plastic, allowing the weeds to grow. And although they will most likely not break through the plastic (unless it is very weak plastic), you will still have to weed under the plastic, and your vegetables will still be competing for nutrients, both of which will defeat the purpose of having put the plastic down in the first place.

Other mulching options include straw and cut grass. However, when going this route, it may actually create more of a problem than it is meant to prevent. If there are any seeds in the straw or grasses, they will inevitably sprout along the way, making the mulch work against you. These methods can work in a pinch, but you will still need to watch for weed growth.

Although not entirely weed free, raised beds do have much fewer problems with weeds and for gardeners with little time or patience to mulch; this would be the next best thing to no weeds at all. There still may be weeds here and there, as weed seeds may blow in from the wind or be spread by birds and small animals. However, because the bed is not flat soil that is already infested with weed seeds, any weeds that do find their way in will be few, manageable, and easy to pull. As a matter of fact, because raised beds have much less weeding and do hold moisture quite well, even though the initial building of the beds will take some time and a little money, they can be the way to go if you're a gardener with little time for upkeep.

Although it seems like container gardens would be weed free, this is rarely the case. Of course there will not be the same sort of weeding involved as with the traditional garden, but as with the raised beds, weed seeds will find their way into the pots and soil. As with the raised beds, any weed that does find its way into the containers will be easy to remove, making for easy weed control.

Pest Control

In addition to regular watering and weeding, another aspect of garden maintenance involves **pest control.** One problem that gardeners have faced since the beginning of gardening history has been the daily garden raids by the local wildlife as well as insect infestation.

When the local wildlife is using your garden as a snack bar, two options for pest control include fencing and bird netting. Fencing in your garden will keep most ground critters at bay, while bird netting will keep the plants off limits to the birds, without damaging the plants or injuring the birds.

For wildlife, there are other methods besides netting and fencing that can help and are safe for humans, animals, and gardens. Hot-pepper wax, predator urine, and hair scattered around the garden are a few of the most popular examples, as well as companion planting, which was discussed in Chapter 4. Most gardeners have had success in using hot-pepper-based products, usually waxes or sprays, but urine's and hair's results have been mixed. Some gardeners have also had success with owl statues, scarecrows, or aluminum pie tins hung and placed strategically in and around the garden. Another method for pest control is to create a sacrificial garden. These little plots typically consist of something very simple, such as lettuce planted in front of the actual garden. The point of the sacrificial garden is that the animals will eat from this garden, leaving the actual garden alone. This method has worked for some, but not for others. It is recommended that you try different methods to see which will work best for your garden.

When planting the seedlings, there is always a threat of insects eating at the stems, as well as other small animals attacking such as rabbits. This next tip will not help much with the rabbits, but can greatly help with the insects that may want to gnaw at the stems. These little plant savers are nothing more than paper collars, and you (or your kids) can make them easily on their own.

Making newspaper collars to protect your seedlings. Illustration by Ariel Delacroix Dax.

Cut newspapers into rectangles approximately five inches long and three to four inches wide. Fold the rectangle into thirds or in half, depending on how high you want the collar to be, forming a strip. Take this strip and very loosely make a collar around the base of the plant, then fasten using weatherproof tape or even staples. The paper collar may remain at the base of the plant or be removed when you deem it is all right to do so.

For insects, one of the best methods of "bug proofing" the garden is through companion planting. There are a number of lists online as well as books and magazines that will provide the gardener with the right combination of plant companions that will act as pest control for others.

If you keep chickens, you can let them roam through the garden to help with insect control. But again, while this method has worked well for many, others find that their chickens sometimes will peck at some of the fruits or vegetables that are growing as well. For those gardeners who do have chickens, it is worthwhile to give them a try, but watch carefully during the first few tries. If the chickens do work out, they can be some of the best insect control in the garden. And they work for free!

When deer are a problem in the garden, and you don't care to fence the garden in, deer-resistant plantings around the perimeter of the garden may be of assistance. For further information on this type of planting, inquire at nurseries to be directed to the correct plants for their areas.

For slugs and snails, simply scatter pans of beer throughout the garden. The beer attracts the slugs; they crawl in, drink, and drown. Bye-bye, slugs and snails. You can also slightly sink the pans into the soil to keep them from tipping.

Protecting the garden against wildlife and pests can seem like a big job. However, today, there are numerous sources both online and in books or magazines, as well as garden centers, chat groups, and extension offices, making information on protecting the garden from wildlife and pests much easier to obtain and figure out. But it is still up to you to decide which is the correct method (or methods) that will work in your garden and fill your needs.

Dealing with Cold and Frost

For those fortunate enough to be gardening in a mild or warm climate, extreme cold or frost is usually not an issue. However, many gardens are in areas where extreme cold and frost is an issue. Serious gardeners in these cold weather areas find that garden blankets, cold frames, tunnels, and greenhouses (heated or not) can help extend the growing season, both at the beginning and end.

These seasonal extenders can be found in a variety of sizes and price ranges, as well as differences in quality. Many farm and seed retailers carry these seasonal extenders in their stores, on their websites, and in catalogs. Some home stores and nurseries may carry a limited inventory as well. And for the "do-it-yourself" crowd, the Web, garden books, and homesteading and garden magazines are full of information for how-tos in constructing cold frames and hoop houses and tunnels, as well as providing materials lists.

Garden Blankets

Garden blankets are simply cloths made for the garden that you can cover and protect your plants with to help combat the threat of frost. Most garden centers will have them for sale. Just put them on the plant for the night to protect from frost damage.

If garden blankets cannot be located or are just not in your budget, bed sheets are a great substitution. They are lightweight, so they won't crush the plants, and are large enough to cover a generous space. If bed sheets are not available, lightweight bath towels will work as a backup. However, one type of covering that should never be used, and this cannot be stressed enough, is plastic sheeting of any kind. When covering plants in preparation for a frost, instead of protecting the plants, the plastic will actually encourage and allow ice to form underneath, freezing the plants instead of protecting them. As a result, the plants will be lost.

Cold Frames

Cold frames are glass-covered outdoor boxes that extend the growing season at both ends. In the early spring, cold frames may be used to harden off seedlings that were started indoors. This means that the seedlings are being allowed to adjust to the outdoor temperature while still being protected against the cool nights of early spring (and last-minute frosts) through the raising and lowering of the lid, allowing the temperature adjustment.

At the other end of the seasonal spectrum, some greens, like leaf lettuce, mescaline, and spinach, may be grown in the colder time of the year in the cold frame. Even though

A medium-sized cold frame. Photo courtesy of PALRAM Applications Ltd. (www.palramapplications.com)

the mentioned greens arc cold hearty, they will usually not survive a killing frost. But planted in a cold frame, they are protected from frosts, allowing the greens to continue growing into at least early winter.

Cold frames can be as simple or as complex to build as you choose, and can be built from recycled materials or newly purchased ones. There are a number of variations, both online and in gardening books and magazines; you just need to do a bit of research to discover which style and size is best for your needs. You may also choose to design your own cold frame as well. Kits are also available. Remember, the cold frames do not need to be fancy in order to work well.

Hoop Houses and Tunnels

Hoop houses and **tunnels** may also extend the growing period of your garden. These structures are nothing more than mini-greenhouses just high enough to cover the plants or gardens, sometimes being high enough for

A large-sized hoop house. Photo courtesy of www.onestrawrob.com

you to stand in, but this is not always the case. They may be purchased or you can purchase the frame ribs and covering materials to build your own. Many larger farm stores will have the materials necessary, but if there is not one nearby, supplies may be ordered through greenhouse- and farm-supply catalogs or websites.

CHAPTER 7

GETTING THE MOST FROM YOUR HARVEST

. . . .

After all the planning, planting, and nurturing, your garden has finally come to fruition, and along with it, an abundance of produce. You've likely grown much more than what can be eaten in a short time, but after all the work, you don't necessarily want to see your bounty spoiling on the vines. This is where food preservation comes in. Canning, freezing, drying, and smoking are just a few ways to preserve your harvest, with canning and freezing being the most popular and most widely utilized with home gardeners.

In the following pages, you will find a quick primer with short explanations of preservation techniques. Before you decide on which method to use, it is recommended that you read in depth about each for hints, tips, and dos and don'ts.

Freezing

Freezing is perhaps the one form of food preservation that almost everyone has used at some point, gardeners and non-gardeners alike. Like any other preservation technique, freezing has pros and cons. Perhaps the best "pro" is that anyone of any age can do it. In addition, except for either freezer bags or containers, there is no special equipment required for freezing (unless vacuum packaging is to be used, in which case home equipment may be purchased). Prep work for freezing is also fairly easy.

The prep work involved will depend on what is to be frozen. Some foods, such as beans, peas, and carrots, may need to be blanched first with a quick dip in boiling water. Some vegetables or fruits may also need peeling, chopping, or pureeing. It is all up to you as head gardener and cook.

One drawback of freezing, especially when trying to freeze small, individual pieces in one bag (such as berries), is the food freezing together in one huge clump. This can, however, be easily remedied. When freezing small, loose items like peas, carrots (whole or sliced), berries, or anything else that could freeze together into one large rock formation if just tossed into a bag, spread the items out on a cookie sheet covered with either waxed paper or parchment paper. Try to make sure the items are not touching each other. Then place the entire tray into the freezer just long enough so the food pieces have individually frozen. Remove the tray. Do not let thaw. Put food into either freezer bags or freezer containers, and return to freezer. Using this method, the food freezes loosely, will not stick together, and you can take out only what is needed, without having to thaw the entire bag or container out first. Not only a neat little trick, but it also saves needless thawing and refreezing, which doesn't always work out well for the food, as foods should only be frozen and thawed once. Refreezing may be okay one time, but after that, taste and quality will lessen.

The main drawback of freezing is the risk of freezer burn if not wrapped properly or used quickly enough. The biggest threat is loss of what has been put up should the freezer fail for any reason. (During a power failure, the fuller the freezer and the less the door is opened and closed, the longer the food will stay frozen.) That said, if you are very new to preserving, freezing is the easiest method to use and, frankly, the one that will keep your food the freshest tasting.

Canning

The next method of preservation, which has been experiencing a rebirth in the last few years, is **canning.** These are the glass jars of food that were seen on many a grandmother's pantry shelf when fall came along. While this is a preservation method that does not require freezing or refrigeration (until the jar is opened), and the filled jars may be stored on the shelf until needed, canning can also pose a few dangers if done improperly. The following is a brief overview of the canning process to give you an idea of what is involved in the work. If you are interested in canning, but have never done it before, carefully follow these step-by-step directions.

Canning is the heat packing of food into a glass container. Unlike freezing, canning does require special equipment for processing, and the initial start-up can be expensive. However, most will be a onetime-only purchase, and can be reused year after year. Many times, useable jars can even be found at garage sales, yard sales, and even sometimes set out free for the taking.

The following is a list of the most common equipment needed for canning:

- Jars
- Lids and rings
- Large pot, to hold jars upright for boiling
- Lint-free towels

The most obvious items needed for canning are the **jars.** Made of glass, canning jars range in size from half pint to one quart. The jar openings or "mouths" may be large or small. The jars may be used year after year, but must be carefully examined for chips and cracks before each use. If any impurities are found, the jar must be discarded. Note: it is important to use only jars made for canning, as they must be able to stand up to repeated boiling.

Pickling green beans using common canning supplies.

Along with the jars, **lids** and **rings** will also be needed. The lids and rings make up the cap for the jar. As with the jar mouths, lids and rings come in various sizes. While the rings may be used year to year, the lids are made for one-time use only and should be Before each use, whether the jars are new or used, they need to be sterilized and inspected for chips and cracks. Chipping can especially occur around the rim of the mouth of the jar, so pay close attention to this area. After passing inspection, the jars should be thoroughly washed with soap and water, and then carefully put in a pot of boiling water for at least ten minutes to sterilize. Canning pots are available in most stores, but any pot that is large and deep enough for your needs will do. Although the jars can be carefully set into the pots, there are canning baskets available that can take some of the hazard out of this job. The basket is simply a wire basket that holds four to six jars. The basket is then set into the boiling pot. When the jars are ready to be removed, carefully take hold of the basket handle (using a pot holder) and lift out of the pot.

If a basket is not used, then carefully remove the jars with a set of tongs made especially for grasping canning jars. They are wide and are built to grip around the necks of the jars securely. When removing the jars from the boiling water, whether the basket or tongs are used, it is imperative that caution is used, as the jars will be filled with boiling water, which can carefully be poured back into the pot. Allow the jars to air dry upside down on (preferably) lint-free towels. As the jars dry, they will most likely have a heavy, white film. This is normal and does not need to be wiped or washed off. The lids and rings also need to be sterilized and dried in the same way. Do not towel dry jars, lids, or rings! A dirty towel may contaminate the jars or leave lint in or on the jar, causing further contamination.

After jars are properly prepped, the foods may be added in whatever way you have decided to prepare and preserve them. They can be whole (if vegetables like tomatoes, beans, potatoes, beets, or anything else small), sliced, cubed, pureed, or pickled. It all depends on what you choose to do with what you have and how you like to eat it. After the jars are filled, a wooden skewer or a knife should be poked along the inside of the filled jar around the entire perimeter to release any air bubbles. Once that is completed, the rim should be cleaned with a damp, lint-free cloth. After this step, the lids and rings may be put on the jars and returned to the boiling bath (you may use a canning basket if one is available). Bath time for the jars can range from five to twenty-five minutes on average. Sometimes, it can be longer, but it all depends on the food and current conditions of the particular canning time. Water height should be a couple inches above the top of the jar.

When the jars are removed from the bath, carefully sit them on a towel-covered countertop, then cover the jars with another clean towel (these may be bath towels, as there does not have to be as much caution with lint at this point). After a short time, there should be a "popping" sound. This is the jars sealing. It should only take a few hours for the jars to complete popping. To tell if the jars are properly sealed, touch the lid. If the lid gives when touched, the jar is not sealed and should be returned to the bath. If after it is returned to the bath, the jar still does not seal, the contents should either be refrigerated and used or poured into a freezer bag and frozen. (Freezing will not work with pickling, so anything pickled will need to be refrigerated.)

Properly sealed jars may sit safely on the shelf with no cooling necessary. However, the jars should occasionally be checked for their seal. If the lid is no longer tight and gives when touched, the jar contents and lid should be discarded. The jar and ring may be salvaged and thoroughly washed to be used again (as always, check the jar for chips and cracks.)

For the most part, freezing or canning is your preference. Obviously, canning takes more time than freezing, so if time is an issue, then that would be the most practical way to go. However, some things can better than freeze, and freeze better than can. For example, most greens, like kale, collards, and spinach, should be frozen. For best results, broccoli and cauliflower should be frozen as well. Anything pickled, however, must be canned.

Drying

Drying is another useful method of preservation. Although not quite as popular as freezing or canning, drying is starting to be used more and more. Most herbs, vegetables, and flowers may be dried, with vegetables usually being rehydrated before consumption or use. Rehydration simply involves placing the dried food in liquid until it plumps up once again and softens.

Herbs and flowers do well with air-drying, while vegetables do best in a dehydrator or in a low and slow oven. Thoroughly dried items probably have the longest shelf life and can be stored in jars, plastic containers, or plastic bags. It is important, however, that no moisture get into the container, as this will mold the dried Air-drying is as simple as tying the plants together and hanging them in a room with little dust exposure and out of direct light, as light exposure could fade the color of the foods.

Electric dehydrators will allow sliced vegetables and fruits to dry on shelves within it. The dehydrator can sit anywhere you have an outlet. Some dehydrators will work better than others, and you may experience uneven drying. If this happens, you may need to periodically change the position of the shelves inside of the dehydrator to allow even drying. Dehydrators are available in an assortment of sizes and price ranges, and are easy to find, as so many households are using them today.

Drying may also be done in low, slow ovens, no higher than 125 to 200°F. Not very energy efficient, however, if you are concerned about power usage, unless you are drying a large amount at one time. If you plan on doing a lot of drying of vegetables (and fruit), I suggest that you do your research and purchase the best dehydrator that meets your needs and that your budget will allow.

Market the Excess

If after preserving your harvest and sharing it with family and friends, your garden still has an abundance of produce left, this may be the time to let the garden earn you some extra money. There are a few ways to do this, although the two

Fresh produce off to the market. Photo courtesy of Catarina Astrom (www.catastrom.com)

most common ways are through a roadside stand or a farmer's market. The direction you take will depend on how much produce you have to sell and how much time you have to put into sales.

Setting up a roadside stand can be as easy as setting up a bench with neatly stacked baskets of vegetables or as complex as building a small stand with shelves. Before you set up a stand, be sure to check your local zoning and permit regulations. Most areas do not require permits for this, but it is always better to be safe. It is best to call your town offices to find out.

If you have an abundance of produce, a booth at your local **farmer's market** may also be a consideration. These markets are a wonderful place not only for sales, but for networking with other food producers, large and small. However, most farmer's markets do have booth fees and insurance is required, so you will need to be sure that the expense will be worthwhile.

For those who do not want to take the time to sell their food, you can find a local farm market that will sell your produce in their store, or look for another gardener or farmer who will take it to the farmer's market and sell it in their booth for a percentage of the sale or sharing booth costs.

No matter how you decide to sell your extra produce, there are a few important things to consider before getting started. First, make sure your produce is clean and has no rot spots or worms. While most consumers will not fuss about a few specs of dirt or some dust from freshly harvested vegetables, produce encrusted with dirt or not in good condition will turn the consumer off to the product being offered. Also, be sure to keep fresh cut herbs looking fresh if there is a lapse of time between picking and market time (but no more than a few hours). Place the herbs in containers full of water and put in the refrigerator until ready to leave for the market. Once at the market, the herbs may be displayed in the same containers of water to keep them looking appealing for customers. The herbs should also be bunched with each bunch held together with a small rubber band to allow the consumer to easily remove from the container to purchase.

Second, take the time to create a nice-looking display. This doesn't mean that it needs to be fancy, but it should be made to look appealing and inviting. Even a bench in your front yard loaded with vegetables can be eye catching simply with a nice arrangement and a crate or two.

Especially when selling at farmer's markets, make an effort to be personable and to interact with your customers. Be prepared to answer questions about how to use a vegetable, fruit, or herb if it is unusual. And if you are offering an unusual item, provide a recipe for it. Your customers are more likely to try something new if they have a way to use it. Make sure that you have the recipes sitting next to the item, as this will encourage your customers to make the purchase. You may be able to create a profitable niche market for yourself along the way if you become known for unique foods. It also wouldn't hurt to have little blurbs about some common or uncommon items, saying things such as "delicious eaten raw," "great for stir-fry," "try in a salad," and so on to give your customers some ideas. Also, be prepared for those who just want to chat for a few minutes. Finally, it's always helpful to show pride in what you produce, and to present the best product possible in the best way possible.

Saving Seeds

One perk of growing heirloom plants (see page 36) is the ability to save seeds for the following season. This is a relatively easy process, but can be a bit time consuming.

When saving seeds from the harvest, make sure the seeds you are taking are from the nonhybrid fruits and vegetables. This simply means saving the seeds from those fruits and vegetables that are not crossbred. To illustrate this in the most basic of examples, it is like the difference between an animal that is purebred and one that is of mixed breed. The offspring of the purebred animals will be exactly what is expected, which would be a copy of the parents.

However, when mixed-breed animals are bred, what the offspring will look like can be anybody's guess. There is a fifty-fifty chance that the offspring will look like one of the parents. And some mixed breeds are sterile, meaning nothing will result.

It is basically (very basically) the same with heirloom (nonhybrid) and hybrid plants. It can be a gamble to save seeds from hybrids. Once planted, the seed may produce a plant that fruits, it may produce a plant only, or it could produce nothing at all.

As a result, hybrid seeds may not be worth the time to save. For those who like to experiment and would like to try saving hybrids, remember that the seeds may or may not produce. But trying a few won't hurt, and could have some interesting results, as well as being fun to try.

When I experimented with hybrid cabbage seeds that I had saved from the previous year, after planting five seeds, my results were as follows:

- Two seeds produced cabbages. Although they were much smaller than usual for that particular type, they were edible and had good taste.
- Two seeds produced plants, but no cabbage.
- One seed did not germinate at all.

Although I did have fun with the experiment, and would try it again with another vegetable type, the results were not good enough to justify saving seeds from hybrids on a continual basis for the majority of the following year's planting.

So what exactly does the process of seed-saving entail? The first thing is to select the source of the seed. Where the seed will come from depends on the item. Sometimes, like with herbs or lettuce, the seed comes from the plant when flowers form. Other times, they will come from the fruit or vegetable, such as apples, cucumbers, tomatoes, and squash.

To harvest seeds from plants, the plants first need to go to seed, meaning there needs to be some plants selected that will not be harvested at a certain point, but instead, will be left to grow out allowing flowers and seeds to form. The type of plant will dictate how the seeds are harvested. Some must be allowed to dry on the plant, while other plants may be cut just before the seeds are ready to take, and then you suspend the plant/seed heads in a small brown bag, allowing the seeds to drop inside the bag, making collection much easier.

For harvesting seeds from a fruit or vegetable, most of the time it is as easy as cutting the item in half (usually top to bottom), carefully removing the seeds, then drying and packaging. Others, like tomatoes, have a little more of a process for harvesting their seeds, but this is still not complicated. All seeds harvested should be stored in a labeled envelope in a dry, dark, cool place.

This has been only a very brief introduction to seed-saving. There are many resources online, in books, and in magazines dedicated to the art of seed-saving, including how to select the best vegetable to pull seeds from (usually, the best examples of the fruit that a plant produces should be the one saved for seed), extraction techniques, drying hints and tips, and proper storage, and many will provide information on seed exchange or swap groups. For the gardener who is serious about saving their own seeds year to year, it is worthwhile to invest in a good manual on the topic.

Harvesting and saving seeds is a fun activity and can also be educational for kids. It provides quite a sense of accomplishment when the seeds planted next growing season are the seeds that were harvested from the garden the last growing season. And if there are seeds left over or an abundance of a few types of seed, a seed swap is a fun and inexpensive way to obtain new and different seeds for your garden, not to mention a good excuse for a group of gardeners to get together, brag, and share information about their gardens and experiences.

CHAPTER 8

END OF THE
SEASON

he harvesting has just about come to an end, and your seeds have been collected, saved, and properly stored. Now, it is time to start cleaning out and buttoning up your garden for the winter. However, before the clean-out begins, it is always wise to double-check for any last-minute items that may have ripened late or may be saved and ripened off the vine.

"Winterizing" in Warmer Climates

If you are lucky enough to live in a region where gardening can take place almost year-round, your garden will continue to flourish. However, plants may need to be swapped out once they reach their peak, due to the fact that even in warm climates, most all-annual vegetable plants will still have a certain lifespan, although it will be longer than in colder climates. If they are a tender perennial, as discussed earlier, they will be a perennial in the warm climate, even though they are an annual in a colder one.

And in many warm climates, instead of gardening time coming to a standstill in the winter, the hot summers may be your time to sit back and relax a bit, as, if you are in an area that can get hot, hot summers, it can be too hot for some plants during that time (time varies with region: usually anywhere from July to September, give or take a few weeks in either direction).

End of Season Harvests

Green tomatoes found after harvest may be picked and brought inside to continue ripening. You can simply set the tomatoes on a windowsill or wrap them individually in newspaper, lay them carefully in a box, and store in a cool place to ripen slowly. There are also a number of recipes in which green tomatoes are the star attraction.

Root crops like carrots and turnips, can even be stored right in the garden. Just leave in place and cover with a layer of straw at least six inches thick. During the winter when you want to harvest, simply lift the straw, pull or dig out what is needed, and then put the straw back in place to cover whatever else is left.

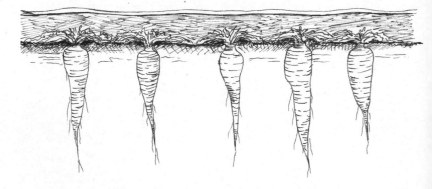

Root crops, such as carrots, can be covered with straw and left in the garden. Illustration by Ariel Delacroix Dax.

If herbs or other trees or plants are to be overwintered as well, it is necessary to make sure that they are properly protected, if they need extra protection. Not every plant or tree will need extra help over the winter, so if you are unsure about the plants to be overwintered, it would be best to do a little research and inquiry about the plant's needs. And if you would prefer not to deal with "special-needs" plants in the winter, then I suggest that any plants that you are unsure of be further researched before the final selections are made.

Because root crops left in the ground and harvested throughout the winter have had time for sugars to really form, they often taste sweet, crisp, and delicious. They can even taste better than those vegetables from the summer and fall harvest.

Root crops may also be stored in root cellars and cold rooms. These storage areas are simply cellars or unheated rooms within a home that, although the temperatures do not get to freezing, they stay cold enough to keep vegetables usable into the winter months without having to keep in a refrigerator. These storage areas still need to be monitored, as the stored food will not keep indefinitely, even in the best situation.

Cover Crops

Before finally putting the garden to bed for the winter, some gardeners may choose to plant a cover crop for the winter. Cover crops will help fight weeds and add nutrients to the garden. They may be tilled under at the end of the season. There are a number of different crops that the gardener may choose from, mostly depending on location, climate, location of garden, and more. For best results, discuss your needs with a local extension agent for advice and assistance in what the right cover crop will be for your situation.

Putting the Garden to Bed

Once you have saved any leftovers, it is now house-cleaning time. While any perennials will, of course, stay in the garden, as will any root crops that have been properly covered, anything else should be pulled out, roots and all. The plants will usually come out easily, but if they don't, they can be left in the ground, as the plants will not come back next year.

After pulling your plants from the garden, they should not be thrown in the garbage. If you keep any livestock, and the plants have not been covered with pesticides, feed the plants to your animals (just make sure none of the plants are poisonous to them). Don't try saving time by just turning the animals loose in the garden, or it could cause more problems than it solves. The animals will eat anything they see, whether they are supposed to or not, and next season, there may be problems keeping them out of the garden, even if fenced. Always take the plants to them.

If you do not keep livestock, and don't have any friends or neighbors who keep animals, then plants should be composted.

Once the garden has been cleaned out, the soil can be raked up and smoothed out. To help keep the weeds out for the winter, if the garden is quite small, a tarp can be laid over it to inhibit weed growth in the spring (making life so much easier). If the garden is quite large, then use of a tarp would not be practical, in which case the area may be weeded and tilled in the spring, as usual.

Any containers in use for planting should also be taken care of for the winter. Perennials need to be moved indoors or into a heated greenhouse if you have one. Also at this time, any other plants that are still producing may be moved indoors or to the greenhouse as well, to lengthen their growing and producing time. If there is no space to allow for this, then anything that can be harvested should, after which the plants need to be pulled. Once the plants are gone, the pots, especially terra cotta or ceramic ones, should be stored indoors so they don't crack over the winter (this is mostly for areas with ice and snow, although a heavy or killing frost could possibly cause some pot damage, as well, on rare occasions).

Pots may be stored with or without the soil in them. It should be noted that the soil will need to be refreshed or replaced before planting again. Pots should also be washed before the next use if there has been any question about the plant it contained having some sort of disease or fungus.

Manure and Composting

If you are doing organic gardening, manure is used quite a bit as fertilizer for the garden, usually in the form of compost (also known as "black gold" in organic circles). Most manure cannot be put down raw directly onto a planted garden, simply because the raw manure can be too strong and may burn the plants.

Container composting. Photo by mjmonty under the Creative Commons Attribution License 2.0.

When making compost, the pile may be in an out-of-the-way corner of the yard or placed in a bin to "cook." The method you select will most likely depend on space, location, local laws (unfortunately, some areas still don't acknowledge the benefits of the compost pile), and budget.

A good compost pile will have a smell like a forest floor. If a pile smells like rotting garbage, then there is a problem that needs to be addressed. There can be a few reasons that the compost isn't right, and there are many resources available to help with troubleshooting, including the local extension office and many sites online that address composting and compost health.

If red worms have been discovered in the compost pile, this means that the pile is definitely on the right track, and you should keep on doing what you are doing with it. If more worms are desired, additional red worms may be purchased.

Supports for vertical gardens may or may not be removed, depending on whether or not the supports will withstand the harsh winter weather. Removal of the supports would be up to you.

Once the garden is tucked in and the equipment is put away for the year, you can now enjoy a bit of rest and recuperation before it is time to order the new catalogs and begin next year's or next season's wish list. Then the cycle begins once again, with you waiting impatiently for the moment when the seeds can come out to be started, the garden can be opened up and turned if need be, and planting will commence once again.

FINAL NOTES

I hope that you have found this basic primer to be of use in your initial introduction into the world of gardening and home food production.

As stated within the book, local extension agents and greenhouses can be an invaluable source of information to any new gardener, as is the Internet. However, the best way to keep on learning is to network with other gardeners, join a local or online garden club, and frequent local nurseries.

Finally, keep on reading and don't be afraid to experiment, maybe even failing once or twice. Gardening is a lot like on-the-job training...but a lot more fun!

Enjoy!

RESOURCES

USDA.gov
To navigate the entire USDA site, this is the easiest way to begin:

At the home page (http://www.usda.gov), look at the links at the top right of the page. Select either "A-Z Index" or "Site Map." Either of these selections will assist with easier navigation of the USDA website. There are also search boxes, usually in the left hand corner, that may be of some assistance if you type in what you are looking for.

USDA Extension Offices (http://www.csrees. usda.gov/Extension)
This convenient website will allow you to quickly and easily locate the extension offices in your area. Extension offices serve as an invaluable resource to new gardeners, as they keep track of local weather, soil composition, gardening regulations and restrictions, and so much more.

USDA Plant Hardiness Map (http:// planthardiness.ars.usda.gov/PHZMWeb)
This map provides an excellent resource for new gardeners to better understand the environment and growing conditions in their state and town. It may be printed out for home use.

USDA Organic Certification (www.usda.gov/ wps/portal/usda/usdahome?navid=ORGANIC_ CERTIFICATIO)
This section of the USDA website provides a more in-depth look at the requirements to become certified as an organic grower. Containing both general information for those who are looking to start and specific information for states and counties, any gardener interested in a chemical-free garden, either now or in the future, should take the time to read this site.

SEED CATALOGS:

John Scheepers Kitchen Seeds (http://www. kitchenseeds.com)
In business since 1908, this beautifully laid-out and illustrated catalog is a joy to look through, even for the nongardener. For the novice gardener, the catalog contains simple, easy-to-care-for plants, as well as new varieties and hybrids for the more adventurous types.

The Cook's Garden (http://www.cooksgarden.com)
A newer company, this catalog's niche is seeds for the gourmet gardener. Containing tried-and-tested new varieties of old breeds, these seeds produce some of the most flavorful produce around. For slightly more experienced gardeners looking to expand their variety and their palate, this catalog is an excellent choice.

Burpee (http://www.burpee.com)
A great resource for both old favorites and brand-new experimental hybrids, the curiously named Burpee provides a great resource for a third- or fourth-year gardener who is looking to increase the breadth of his garden's variety or who is starting to think about trying to make their own hybrids.

Baker Creek (http://www.rareseeds.com)
A must for anyone interested in growing heirlooms in their gardens. Chock-full of familiar and unusual vegetables (and fruits) to grow, as well as excellent photographs of the vegetables (or fruits), this is one of the best resources out there.

Seeds of Change (http://www.seedsofchange.com)
This is a purveyor of certified organic seeds. Although they may now be found in various brick-and-mortar market venues, in order to really see the depth of their offerings, those serious in organics need to take a look at the entire catalog.

Seed Savers Exchange (http://www.seedsavers.org)
Dedicated to saving and sharing heirloom seeds, this is another excellent resource for the usual and the unusual for the food garden. Customers may also become members of this organization and become stewards in the movement to save and spread heirlooms.

WEBSITES:

Garden.org
An excellent resource for any gardener, the website of the National Gardening Association provides up-to-date information on gardening, including tips and tricks for beginners, as well as a variety of other resources, including an easy-to-navigate "Gardening Dictionary" to aid in your planning and planting.

Gardeners.com

In addition to a valuable, reasonably priced online catalog, the website of the Gardener's Supply Company also provides you with resources like expert opinions, helpful tips, and simple, affordable tools that will make your garden more fruitful, while being easier to manage.

Davesgarden.com

Dave's Garden website is built with new gardeners in mind. The plant database is one of the largest in the world with over 80,000 entries. You can search by name or plant characteristics, or just browse through the pictures. Searches are limited if you are not a registered member, but membership is free.

BOOKS/MAGAZINES:

The Vegetable Gardener's Bible by Edward C. Smith

This is one of the best reference texts you can have on hand when planning out your vegetable garden. With extremely in-depth looks at a wide variety of vegetables, what goes into growing them, and how to make sure they flourish, this book will serve as an excellent supplement to Gardening Made Easy.

Rodale's Illustrated Encyclopedia of Organic Gardening by Anna Kruger

This is an excellent starting point for new gardeners interested in growing organic, but unsure of where to start and concerned about potential difficulties. Written in simple terms with beautiful, full-color illustrations, this book will let you put your best food forward in growing organic.

Let it Rot!: The Gardener's Guide to Composting by Stu Campbell

A tried-and-true resource for the gardening community, the third edition of *Let it Rot!* includes additional tips on how to start composting, how to maintain a compost heap, and how best to make use of compost to nourish and enrich your soil. While composting isn't a resource available to all new gardeners, this book will help those with sufficient space to get started.

The Complete Guide to Companion Planting: Everything You Need to Know to Make Your Garden Successful by Dale Mayer

Companion planting can be one of the most rewarding methods of improving your garden's output and making the entire experience simple and efficient. This is the best resource to have on hand when planning your garden, both to maximize development and to make sure that your plant's neighbors aren't going to be impeding their growth.

Fine Gardening

This magazine not only provides a wealth of information online (http://www.finegardening.com), but also provides the option for a monthly subscription. In addition to providing the new gardener with valuable resources and information on gardening and raising plants, it also provides a wealth of information on the practical aspects of landscaping, helping those who want their garden to be a visual centerpiece as well as a food source.

Mother Earth News

One of the originals, this magazine covers everything from self-sustaining living to the use of alternative energies. The magazine is loaded with information on gardens, animal rearing and husbandry, homesteading how-to's, and much more.

Organic Gardening

As the title states, this magazine focuses on organic gardening through hints, tips, methods, recipes, and wonderful photography. The magazine is a great addition to the library of anyone interested in growing organically, whether for home use or market.

Countryside

Probably the oldest magazine dedicated to the homestead movement, *Countryside* focuses on home food production. From gardens and meats to preservation and recipes and even the business side of the homestead, *Countryside* continues to serve those who strive to produce their own foods and work to become self-reliant.

Urban Farm

Although geared to today's urban farmer, this magazine is full of information for both the urban farmer and the small rural backyard farmer. From rooftop gardens to small livestock in the backyard, *Urban Farm* is both informative and enjoyable to read.